A
POLAR
AFFAIR

Antarctica's Forgotten Hero and the Secret Love Lives of Penguins

LLOYD SPENCER DAVIS

PEGASUS BOOKS
NEW YORK LONDON

A POLAR AFFAIR

Pegasus Books Ltd.
148 W 37th Street, 13th Floor
New York, NY 10018

ISBN: 978-1-64313-125-2

10 9 8 7 6 5 4 3 2 1

Printed in the United States of America
Distributed by W. W. Norton & Company, Inc.
www.pegasusbooks.us

To Wiebke

for completing my song

CONTENTS

PROLOGUE

I t is October 13, 1911. A cold day in Antarctica, even by Antarctic standards. Pack ice extends to the horizon from the tiny spit of land that is Ridley Beach at Cape Adare. A white, windswept tableau, it has been battered into a state of bleakness by the blizzard that masses overhead and pushes hard up against the mountains it hides. It is as uninviting a place as it is possible to find. And yet, as the man squints ahead of him, he can make out a little black-and-white person, a penguin, leaning into the wind, straining, marching toward its destination. This beach. This season. A chance to prove itself.

The man looks at the penguin with some ambivalence. He would rather be pulling sledges than stuck here, marooned on a piece of land not much bigger than a football field, which he will soon have to share with thousands of the little monochromatic blighters that are somewhere out there, in the snow and the ice, moving unerringly toward him. He is dressed against the cold, but his face remains exposed. Still tanned by the sun even after a winter's darkness, you could be forgiven for mistaking him as someone other than an Englishman. That night, in the comparative warmth of his hut, with a sigh that is part frustration, part fascination, he takes a fountain pen and in a blue-bound notebook writes, "1st Penguin arrives Oct. 13th," which he underlines with a sweep of the blue-black ink. It is the beginning of the world's first serious study of penguins.

It is January 14, 1912. A cold day in Antarctica, even by Antarctic standards. A line of men, real men, strain and lean into the wind as they pull a laden sledge across an ice-covered flat and featureless landscape, trudging toward their destination. They are on a mission to prove themselves. The South Pole. Their destiny. Their duty. As Englishmen.

It is March 29, 1912. Nearly eleven weeks have elapsed. Of the five men who marched to the Pole, only three remain, pressed against each other in their caribou sleeping bags. Outside, the wind tears mercilessly at their tent, the spindrift of snow flying inside the tent taunts them as much as the cravings from their empty bellies. To go outside means instant death; to stay inside means a lingering one. The Englishman with the round face picks up his pencil. He glances at the still bodies of his companions. Perhaps they are already dead? And, in a small black-bound notebook, he writes for the last time.

> We shall stick it out to the end, but we are getting weaker, of course, and the end cannot be far. It seems a pity, but I do not think I can write more.

Those words would help turn Robert Falcon Scott, failed polar explorer, into a hero: a household name synonymous with courage and perseverance against the odds; a man embraced by a nation. Even his dying postscript—*For God's sake look after our people*—sounds more like a plea to a higher authority to go easy on the whole populace of Great Britain than it does the last wish of a dying man for his beloved wife and son.

Scott's frozen body lies half in, half out of his sleeping bag; an arm over his good friend, the doctor Edward Wilson; his little black notebook and pencil tucked beneath his shoulder. The scrawled signature: R. Scott. Not Robert or Robert Scott, but R. Scott. Stiff and formal, British through and through, even in death.

It is the same day, March 29, 1912, and 220 miles to the north of Scott's tent, the other Englishman, the one who had been studying penguins—also a doctor, also a member of Scott's last Antarctic expedition—crawls into a sleeping bag in a hole dug into a bank of snow where he shall spend the entire Antarctic winter with five companions. He picks up his own pencil and writes, "Blowing hard all day." It is the beginning of the most amazing story of Antarctic survival and adventure ever told, in a place replete with amazing stories and the frozen last breaths of other adventurers. It is the story of the unsung hero of Scott's Terra Nova Expedition, a man who would become the world's first penguin biologist but whose achievements will be lost in the public adulation and sympathy for Scott and his thwarted attainments.

It is the story of George Murray Levick.

PART ONE

THE LURE OF ANTARCTICA

Homosexuality

There is an impression that nature is not just red in tooth and claw, but red-blooded to boot, especially when it comes to sex. Males compete with each other to fornicate with as many females as they can. It is why male elephant seals are so much bigger than their female counterparts; it is why male red deer have antlers; it is, after all, the basis for evolution according to Darwin. *Survival of the fittest* is really a euphemism for *survival of those who fuck the most* and thereby sire the most offspring. Survival, in this sense, has got very little to do with longevity: you survive through your children and their children and so on. Heterosexuality, sex between a male and a female, is not a lifestyle choice so much as it is the essence of life itself, the means by which we perpetuate ourselves—at least as far as we vertebrates know it. Consequently, the notion that sex might occur between members of the same sex has long been seen as unnatural and, given the way natural selection works, unlikely: sex without the prospect of producing progeny would seem to be the epitome of an unsuccessful evolutionary strategy. Homosexuality, using this logic, must be an affectation, a foible of human sociology rather than a product of our biology. For a boy brought up in the Victorian era of Great Britain, sodomy was something sordid, an unspoken evil, and certainly the last thing he would expect to see in wild animals. However, life would make a habit of unfolding in unexpected ways for George Murray Levick.

CHAPTER ONE

VICTORIAN VALUES

I t is October 28, 1996, a good day in Antarctica by Antarctic standards. Clouds cover half the sky, yet visibility is unimpeded all the
way across the sound to massive mountains that sit on the distant
horizon like lost children of the Himalayas. It is a balmy 14°F and the
breeze from the north is not strong enough even to ruffle the patches of
dark blue water that lie sandwiched between pieces of pack ice, which
extend like a giant jigsaw puzzle all the way to the mountains.

I am sitting with a group of Adelie* penguins at the Cape Bird colony
that is located on Ross Island. It is a half-hour helicopter ride from the
hut at Cape Evans where Captain Robert Falcon Scott and his party had
set out on their ultimately fatal mission to become the first humans to
stand on the South Pole. The risks of my own adventure are ostensibly

---

* Adelie penguins were discovered by French explorer Jules Dumont d'Urville, who
named them Adélie penguins after his wife, Adéle. However, in the English-
speaking literature, the accent has been used inconsistently. In 1992, at the
Second International Penguin Conference, penguin researchers agreed on
standardized common names for the various species of penguins. For Adelie
penguins, we settled on Adelie, without the accent.

more those of boredom than any loss of life or limb: I am taking my turn, sitting for hours in the Antarctic open, keeping the mating behavior of the penguins that surround me under constant surveillance. The research being undertaken by me and my team is beginning to reveal that the bedroom antics of the penguins are not exactly like their public personas, which would paint them as virtuous imitations of little people that mate for life; the darlings of the Christian Right; the poster children for monogamy. Not entirely boring, I suppose, but I am certainly not prepared for what follows.

One penguin approaches another. They bow deeply to each other in what is usually a surefire prelude to courtship. Except that, in this instance, both the penguins are males. The approaching male then mounts the other male. If such unexpected debauchery were not surprising enough, afterward, the penguin that has played "female," reciprocates by mounting the first male, ejaculating and depositing sperm on his homosexual lover's genitals in precisely the same manner as occurs in a normal mating between a male and a female penguin.

On this day, I, Lloyd Spencer Davis, penguin biologist, discover something new about the lives of penguins that is completely at odds with the view of penguins promulgated in pretty much every book, documentary, and scientific paper, which collectively suggest that penguins are prim and proper, monogamous little creatures that mate for life, such that if one needs a blueprint against which to measure the human ideals of marriage and fidelity, one need look no further than penguins.

At least, so I thought.

―――

Fifteen years later, Douglas Russell, who goes by the unlikely title of senior curator of birds' eggs and nests at Britain's Natural History Museum, is sifting through a filing box of reprints in the library at the museum's storage and research facilities in Tring. He takes out a three-page printed manuscript that he has never seen before. Printed along its top are the words, NOT FOR PUBLICATION. The manuscript is

entitled, *The Sexual Habits of the Adélie Penguin* and it is written by
Royal Navy Staff Surgeon G. Murray Levick, a doctor who had accompanied Captain Scott to Antarctica and acted as sometime zoologist
and photographer in addition to his medical duties.

Russell has chanced upon seemingly the only surviving copy of a manuscript written by Levick in 1915 about the sexual behavior of penguins—one
of apparently one hundred copies—that got as far as being printed but then,
for whatever reason, was prevented from being published.

Robert Falcon Scott's last expedition to Antarctica, known as the
Terra Nova Expedition, was as much a scientific quest as it had been
a quest to get to the South Pole. At its conclusion, it was incumbent
upon the surviving members of the expedition to publish the results
of their findings. Murray Levick, who had studied Adelie penguins
while stationed at Cape Adare, duly produced a book called *Antarctic
Penguins: A Study of Their Social Habits*, which was published in 1914.
It was the first book ever published about penguins.

I had come across Levick's book in 1977, when I began my own studies
of Adelie penguins in Antarctica. It was one of three books about penguins
that I took down to the Antarctic with me, along with others by Henry
David Thoreau and Walt Whitman. While, to be sure, it had provided
some baseline behavioral observations of Adelie penguins, it otherwise
had seemed fairly quaint to me in the way it characterized the penguins.

For the next thirty-five years, I went about my merry way, blithely
"discovering" the truth about the sexual behavior of penguins. That
is until, nearly a century after it had originally been written, Douglas
Russell published Levick's paper on the sexual habits of Adelie penguins
in the journal *Polar Record*, along with a commentary by him and two
penguin researchers, Bill Sladen and David Ainley.

It is late in the evening and I am sitting at my office computer
reading Levick's belatedly published paper on my screen. The building
is deserted; everyone has gone home. I should have too, but I am glued
to the seat of my red leatherette chair, unable to go anywhere. In his
paper, Levick describes a litany of sexual depravities and misbehaviors
committed by penguins.

I am stunned. Staggered. Like I have been punched in the guts. It is bewildering, but strangely exciting too. I am struck by the realization that, for the better part of my career as a penguin researcher, I have been merely rediscovering what Levick had already discovered about the sexual behavior of penguins. It is as if George Murray Levick, having been denied his own voice, has somehow channeled himself through me. Nowhere is this clearer than in the last two sentences of Levick's document.

> *Here on one occasion I saw what I took to be a cock copulating with a hen. When he had finished, however, and got off, the apparent hen turned out to be a cock, and the act was again performed with their positions reversed, the original "hen" climbing on to the back of the original cock, whereupon the nature of their proceeding was disclosed.*

It could have been exactly the same encounter I had observed on Cape Bird and then described in my own manuscript, which I had published as original research eighty-three years after Levick's observations!

I lean back in my chair, my hands gripping its chrome arms, and swivel around to glance at the bookshelves behind me. They are laden with books about penguins and filing boxes full of reprints about penguins that have been labeled and sorted alphabetically by the first authors' surnames. There must be copies of over two thousand scientific papers about penguins sitting on those shelves, a veritable compendium of penguin research and all we know about these distinctive and charismatic creatures. And yet, save for a few inconsequential notes from early explorers, there is not a single substantive scientific paper about penguins among them that predates 1915. Arranged along the bottom two shelves there are, if I include those with my name on their dust jackets, forty or so books about penguins. In their midst is the somewhat tattered green cover of my copy of Levick's book, *Antarctic Penguins: A Study of Their Social Habits*, which had been printed in 1914. Nothing else on those shelves even comes close to its age.

George Murray Levick—or Murray Levick, as he preferred—was indisputably the father of penguin biology: the first real penguin biologist, the first person to study penguins in a systematic way. Everything else on those shelves came after him. Yet far from being quaint, the paper I have just finished reading, written in 1915, although not published until 2012, proves that he had discovered things about penguins that the rest of us who adorn those shelves took another one hundred years to discern.

I spin around in my chair like a child on a playground merry-go-round—gleeful but confused, trying to process everything. I stop before my reflection in the darkened window. The white of my hair is what I see most clearly staring back at me. For three and a half decades, I have believed that I have been forging my own path: a scientific explorer intent on exposing the truth about the mating habits of penguins. In fact, all I have been doing, apparently, is following in someone else's footsteps, even if time, or censorship, or whatever, has obliterated his footprints.

To learn that Levick has walked before me like an unseen ghost, well, I realize then and there that I need to get to know this man who has been like my personal Sherpa, if silent and invisible. Why was he silenced, censored, prevented from letting the world know about the truth about penguins? By whom, and for what reason? Or, did he choose to shut up about what he had seen? Douglas Russell's commentary in *Polar Biology* mentions evidence that suggests Levick himself may have been complicit in the silencing of his results: he had covered up the most salacious parts of his field notes with a code that used Greek letters. Why?

I need to know. Levick is an enigma to me. I know he wrote his book about penguins, of course, but I know little else. A quick online search suggests that neither does anyone else. The records about Levick are sparse, at best.

He was born in the English city of Newcastle upon Tyne in 1876. He studied medicine. He joined the Royal Navy. He accompanied Scott on the Terra Nova Expedition and, as a member of the Northern Party, overwintered on Inexpressible Island in a snow cave. He served in the First World War, made a name for himself as a doctor afterward, and established something called the Public Schools Exploring Society. He died in 1956.

As I drive home for dinner, rehearsing my apology for being late, it occurs to me that if I am to turn detective and discover the real story behind Murray Levick, then I should start at the beginning. As it happens, I am due to go to England and a side trip to Newcastle seems like it should be as good a place as any to begin to find out about Levick and what made him the man I would become.

Murray Levick was a child of the Victorian era. Born on July 3, 1876, in Newcastle upon Tyne in the middle of the reign of Queen Victoria, he was brought up in a society where the acquisition of high personal moral standards was a developmental stage every bit as expected and predictable as a child's second teeth. Furthermore, while teeth may fall out over a lifetime, no such lapses were excusable when it came to morals. Sex was something reserved for heterosexual couples and, even then, only if they were married to each other.

It has not always been that way. Newcastle, which occupies a jot of land in the northeast of England, has a history of debauchery every bit as profound as its reputation for being an industrial town, depressed and blackened with coal dust. It is true that the expression "taking coals to Newcastle" originated there and, for more than seven hundred years, Newcastle really was the coal capital of Great Britain, the literal engine room for the Industrial Revolution. However, long before that, it had a reputation for being dirty in another kind of way.

When I arrive there, my first impression of Newcastle is that it is all shiny-bright and ultramodern, nestling comfortably on the banks of the River Tyne with an architectural poem—the Gateshead Millennium Bridge—connecting both sides of the river with glorious sweeping arches. The hotel too is very modern, with views of the river. However, as I look beyond Newcastle upon Tyne's modern veneer, I start to see signs of its ancient roots and a heart that has been beating for hundreds, if not thousands, of years.

The Romans established a fort and small settlement called Pons Aelius on the north bank of the River Tyne at what was then the eastern end of Hadrian's Wall. Aelius was the clan name of Emperor Hadrian, who had visited Britain in 122 C.E. and ordered the construction of a wall. By the beginning of the 5th century, the Romans had put up the shutters and left, leaving the town—by then called Monkchester—to the mercy of first the Danes, and then the Normans. It was William the Conqueror's eldest son, Robert Curthose, who would set the enduring tone for the place and give the town its modern name: he built a castle there in 1080 and, thenceforth, it became known as New Castle.

If ever one needed convincing that it is indeed strange that Murray Levick should have felt compelled to disguise a little bit of bad behavior in penguins behind a veil of Greek letters, one need only look to Robert Curthose and his father for perspective. William the Conqueror, who was also known by the equally prophetic title, William the Bastard—apart from being illegitimate, he was also the instigator of the Domesday Book—was the Duke of Normandy in France until he decided to invade England in 1066 and become its first Norman king. His son Robert would go on to sire several illegitimate children of his own, devastate large areas of France, and fight incessantly with his father and two brothers. But not in the way of normal families: they did so with armies. He once even managed to wound his father in battle. When eventually William was killed, at someone else's hand in yet another battle, Robert succeeded his father as the Duke of Normandy, but it would be his two younger brothers, first William Rufus and then Henry I, who would each become king. A chagrined Robert led several insurrections against his brothers to no avail. After forty some years of raping, pillaging, and warring, Robert, then aged about sixty, was captured by Henry's men and imprisoned for the last quarter of his life.

Sexual misconduct, sibling rivalry, and the inhumane treatment of others were clearly de rigueur in society in the past, to such an extent that the behavior of penguins would have seemed pretty tame by comparison.

Yet the Victorian values of Levick's upbringing, a consequence of the timing of his birth, left him, it seems, unable to even mention masturbation or homosexuality in penguins, let alone in men.

In the 140 years since Murray Levick was born, the pendulum in Newcastle has certainly shifted back the other way, so that the city today is more something that Robert Curthose would recognize rather than the Victorian Levick.

Evening is falling as I make my way down to the River Tyne. The curvaceous Gateshead Millennium Bridge is lit subtly, while on the other side of the river it seems that all the coal in Newcastle is being burnt to light up the modern and equally curvaceous new arts center: it reflects from the river's surface like some sort of giant neon painting. There are lots of people out and about, but I notice that most of the foot traffic that negotiates the sweeping arch of the bridge is moving toward me. At a series of bars on my side of the Tyne, men and women congregate like penguins at the start of the breeding season, seemingly as intent on mating as any of the penguins that Levick and I have studied.

I enter the Pitcher & Piano, a contemporary bar that is all square lines, glass, and aluminum. A young man, in jeans and a T-shirt that barely covers his bulging belly, bumps into me, his attention diverted by a woman wearing a pink dress so short that it would scarcely qualify as a T-shirt. At this and three other bars, I push my way through throngs of bulging bellies and minuscule dresses so that, in breaks between the music, I might question these natives of Newcastle. Not one of them, it transpires, has even heard of Murray Levick.

I cannot help but reflect on the contradictions in all of this. Ever since another Victorian gentleman, Charles Darwin, described the phenomenon of sexual selection, it has been assumed that where males and females in a species look alike, they will be monogamous. Where they look different, like here—with beards and breasts, beer bellies and dresses being just the most obvious manifestations of the many sexual differences on display—then it is likely the species will be polygamous, with successful males having several partners. Yet, as patently sexually different as we are as a species, we live in a world where society's mores, and particularly those emanating from our religious institutions, preach marriage and monogamy for us. Conversely, Levick's observations, and subsequently those of mine, would seem to indicate that penguins, the cartoonists' standby for look-alike conformity, are no more wedded to the idea of monogamy than are the inhabitants of Newcastle that surround me.

Murray Levick's father was George Levick, a civil engineer, and his mother was Jeannie Levick. He had two older sisters, Ruby and Lorna. At the time of his birth, the family lived at 12 Whitworth Place. What the family home might have been like in Levick's day is hard to say: I find that it has been torn down and replaced by a row of adjoining brick look-alike apartments that somehow epitomize the modern Newcastle. It is as if the people of Newcastle, in trying desperately to escape their meaner past, have created a facade that is—with the exception of one bridge—as bland as it is shallow.

Naively, perhaps, I have come to Newcastle expecting it to have some kind of monument to Levick, some traces of his roots. Yet in all of Newcastle I cannot find a single memorial, not a single plaque, with his name. If this were a crime scene, then it is as if all the surfaces have been wiped clean.

I am forced to change direction, to begin anew my quest for Murray Levick. At least there is one place where I know for certain there is evidence that Levick left behind: I must go to find the senior curator of birds' eggs and nests at the Natural History Museum in Tring.

Another line of inquiry also points me to the museum's bird collections: if Levick's own family tree and roots have not proven helpful, perhaps those of the penguins might? Of all the world's nine thousand species of living birds,* just nineteen of them are penguins. The specimens held in collections like those at Tring are more than just an assembly of dead representatives of the living; they also carry with them a history of their interactions with humans.

---

* DNA techniques suggest that there may be more than twice as many species of birds (i.e., over eighteen thousand) compared to those described using differences in their morphology.

Penguins are found only in the Southern Hemisphere, and while they must undoubtedly have been known to natives of South Africa and South America for hundreds if not thousands of years, there are no written records to confirm that. It was not until European explorers ventured far enough south that we have our first confirmed sightings of penguins: in 1497 on Vasco da Gama's rounding of the Cape of Good Hope and in 1520 off the coast of Patagonia during Ferdinand Magellan's famous first circumnavigation of the globe. The African and Magellanic penguins that those early explorers encountered are characterized by having black bands on their white chests and are members of a group of penguins known as the banded penguins. They live at the lowest latitudes of any penguins. Indeed, Galapagos penguins, another member of their group, even live right on the equator. As such, these banded penguins are as far removed as it is possible to be from the Adelie and Emperor penguins encountered by Levick on Captain Scott's Terra Nova Expedition.

Antarctica, however, has no natives. The first sightings by humans of the Antarctic-living Adelie and Emperor penguins could not occur until men were able to sail that far south. And that did not occur until the 19th century, during the life and reign of Queen Victoria from 1837 to 1901. In those days, it was common practice to include a naturalist on voyages of exploration, whose job it was to collect specimens of all the new creatures and plants that would inevitably be encountered. These specimens were then preserved, described, categorized, and typically deposited in museums. Museums like the one in Tring.

A Victorian museum for Victorian discoveries established by people with Victorian values: Could there be a better place to look for Antarctic penguins and my man Levick?

CHAPTER TWO

TERRA AUSTRALIS

E ight years before Murray Levick was born into what were fairly ordinary circumstances in Newcastle, the son of one of Britain's wealthiest families was born in London: the 2nd Baron Rothschild, Lionel Walter Rothschild. Walter, who like Levick preferred his second name to his first, would prove to be not just one of England's richest men but, arguably, its most eccentric too.

He was big but not robust, pathologically shy, and suffered from a speech impediment: a combination of conditions that caused his parents to homeschool him. Except that this was no ordinary home, no row house like those in Newcastle. It was Tring Park Mansion, set in over 3,500 acres of grounds some forty miles to the northwest of London.

The original manor had been listed in William the Conqueror's Domesday Book and, for a while, it was owned by William's grandson, King Stephen of England and his wife Matilda. In the late 17th century, the then secretary of the treasury, Sir Henry Guy, upon being gifted the house and grounds from King Charles II, had a new manor house built that was designed by Sir Christopher Wren, the architect responsible for such iconic buildings as St. Paul's Cathedral. Albeit afterward, Guy was confined to the far less salubrious Tower of London for misappropriating treasury funds to support his predilection for home improvements. Walter's grandfather, Baron Lionel de Rothschild bought Tring Park in 1872 and gifted it to Walter's father, the 1st Baron Rothschild. Hence, from the age of four, Walter lived on an estate the size of a small country, which he populated with weird and wonderful creatures such as kangaroos and exotic birds.

At the age of seven, when Murray Levick was little more than a possibility bouncing around in his mother's womb, Walter Rothschild announced to his parents that when he grew up he wanted to be the owner of a zoological museum. Indeed, Walter proved to be far better at collecting butterflies than he ever was at managing finances in the family's banking businesses. Eventually, his parents relented, and as a twenty-first birthday present built him the Walter Rothschild Zoological Museum on the grounds of Tring Park. It opened to the public in 1892 and would house over two million butterflies, over three hundred thousand bird skins, and a veritable cornucopia of other animals shot, snared, or pickled in that peculiar way Victorians showed their love for nature: by killing it and labeling it.

The grounds at Tring were filled with ever more exotic living creatures too, including emus and rheas. Walter had a particular fondness for zebras and would often be seen riding a carriage pulled by zebras, which he even rode to Buckingham Palace. His penchant for kangaroos remained, with the Australian imports playing havoc with the gardens at Tring, much to the consternation of the gardeners.

Yet there was no escaping the success of the Rothschild Museum either: it housed the largest privately owned zoological collection ever amassed. It also contained one of the best zoological libraries anywhere; one that drew scholars from around the world. In the end, it would be sex—or rather, the Victorian attitude to sex that was so ingrained in both Walter and Murray Levick—that would be the museum's undoing. Walter never married but he had a couple of mistresses. One of them, the wife of a wealthy aristocrat, blackmailed Walter, threatening to make their affair public. As Walter had by then been disinherited by his father, who disapproved of his interest in animals rather than money, he was compelled to sell much of the museum's bird collection to the American Museum of Natural History for $225,000. The message seemingly emanating from Tring and its Victorian-bred occupants was that the public shouldn't get to know about sex, be it among penguins or Barons.

Walter died in 1937, and two years later his nephew, who had inherited Tring Park and the mansion, offered the museum and its remaining collections to the British Museum.

The public face of the British Museum of Natural History, which since 1992 has been called simply the Natural History Museum, is a magnificent cathedral-like building in South Kensington, London. Like Newcastle, it is more facade than fact: much of the museum's vast collections are stored elsewhere, including the bird collections, which are located in fortresslike facilities next to the Rothschild Museum in Tring, sitting on the outer side of the M25 motorway that circles London like a modern-day moat.

As I approach the museum on foot, it becomes clear that the 1st Baron Rothschild had spared no expense on his son's present: the Walter Rothschild Zoological Museum is a beautiful building with an Elizabethan air to it. By contrast, the Natural History Museum's adjoining storage and research facilities look more like bunkers. "Function over form" had presumably been their architect's brief and rightly so: the buildings house, among many other things, 750,000 bird skins from about 95 percent of the world's species of birds, and these all have to be kept in carefully controlled conditions if they are not to deteriorate. They also need to be kept secure.

During the night of June 23, 2009, Edwin Rist, a talented American flautist studying at London's Royal Academy of Music, smashed a window into the museum at Tring and stole 299 bird skins. In addition to music, he had an equal talent tying flies for fishing. He had won silver and bronze medals at the 2006 Irish Open Fly-tying Championships, the equivalent of the Masters at Augusta for those peculiar individuals obsessed with attaching feathers and other bits of glitz to hooks in ways that must seem attractive to themselves, if not to the trout.

While arguably a crime of passion, this was not a spur of the moment thing, but rather carefully planned and premeditated. In the rarified world of fly-tying, the stolen feathers were worth a fortune: "millions of pounds" according to one source. Indeed, after his capture, Rist was required to pay back £125,150, which was the estimated earnings he had derived from selling some of the skins through eBay and the like. The rest he had intended to keep for himself, presumably with gold medals at the next Irish Open in his sights.

If the value and vulnerability of the collections at Tring needed any further underlining, two years later Darren Bennett from Leicester—about seventy miles north of Tring—broke into the museum, smashed a cabinet containing two rhinos, and sawed their horns off. He intended to sell the horns on the black market, where they would have fetched an estimated £240,000. According to newspapers that reported his crime, the value of the rhino horns stemmed largely from their being prized in Asia for medicinal purposes and as aphrodisiacs. Bennett, however, did not realize that the horns, for which he risked so much, were worthless plaster fakes. They had been substituted for the real ones by museum staff worried after Rist's break-in. If that were not bad enough for Bennett, he dropped a glove, which a museum staff member found while biking home. Given that Bennett had, on a previous occasion, troubled the police sufficiently that they should take a blood sample from him, the DNA on his glove was able to be used to identify him more easily than a man with a stomach full of rhino keratin can get an erection.

As a consequence of these two robberies, security to get into the Natural History Museum's buildings at Tring is now extremely tight. When I get to the entrance, I am forced to wait while my credentials are checked and, then, to wait some more while Douglas Russell is fetched. This enforced delay, however, leads to a rather serendipitous discovery as I go to the Rothschild Museum next door to while away my time.

Though the fortified building containing the collections at Tring is not a place for the public—it is more back end than front end for the

Natural History Museum—the original Rothschild Museum is open to the public and contains displays of many of the museum's treasures. At the center of one display, I notice a stuffed Emperor penguin. But this is no ordinary penguin, as Douglas, himself not much taller than the penguin, explains to me after he has been fetched. This Emperor penguin is one of the first encountered by humans. It had been collected by the naturalist Joseph Dalton Hooker during an expedition to Antarctic waters led by British naval officer James Clark Ross. My interest is piqued immediately. Not just because Ross's expedition, from 1839 to 1843, was the first to go so far south, but because it did so in the area where I had studied penguins for so many years: on the appropriately named Ross Island.

Emperor penguins were first described scientifically by the head of ornithology at the British Museum, George Robert Gray: perhaps not from the very specimen in front of me, but certainly from one of those few individuals unlucky enough to have been collected by Hooker on Ross's expedition. Gray gave the species its scientific name of *Aptenodytes forsteri* in honor of a German naturalist who went by the even more magnificent handle, Johann Reinhold Forster. It was Forster who had, at the last minute, replaced Joseph Banks as naturalist on Captain James Cook's second voyage to seek a suspected land mass at the bottom of the world known rather vaguely, and somewhat optimistically, as *Terra Australis Incognita*. This large dollop of southern land had been thought necessary by cartographers to balance the mass of land evident in the Northern Hemisphere.

James Cook and Joseph Banks are very familiar to me, their feats drilled into me during countless lessons at primary and secondary school in New Zealand. It was Cook, a British naval lieutenant, who, in what became known rather unfairly as his First Voyage (he had already been at sea for over twenty years by then) really put New Zealand on the map, so to speak. In 1642, the Dutch explorer Abel Tasman became the first European to see New Zealand, but he did little more than put a few squiggles on a map and leave his name attached to the sea that—some would say, thankfully—separates New Zealand from Australia. It was

Cook who would rediscover New Zealand in 1769, circumnavigate it, and fastidiously map its coastline, while Banks, his onboard naturalist, would collect and describe many of the unique creatures and plant life that make New Zealand much more Middle Earth than any that Tolkien ever wrote about.

Cook, born in 1728, began his life at sea in 1746 when, as an apprentice in the merchant navy, he worked on colliers transporting coal from the Tyne to London, exactly 130 years before the banks of the River Tyne would produce, in addition to coal, one Murray Levick. He volunteered for the Royal Navy in 1755 at the outbreak of the Seven Years' War between Great Britain and France, when Britain decided that it was not going to let France do to its colonies in North America what it had allowed William the Conqueror to do to it in 1066. By 1760, the French had been expelled from Canada and the British were confirmed as the greatest naval power of their day.

Cook had been master of the HMS *Pembroke* that took part in the siege of Quebec. He distinguished himself through his talent for surveying and cartography: he mapped much of the Saint Lawrence River in great detail, and this proved crucial in allowing the British to mount a surprise attack at the Plains of Abraham near Quebec City, the decisive battle and turning point in the Seven Years' War.

James Cook returned to England long enough to get married and then, during five expeditions, he surveyed and mapped the coastline of Newfoundland in what was the first large-scale hydrographic survey to use precise triangulation to establish the outline of land masses. Cook's maps were so accurate that they would be used by those sailing in Newfoundland's waters for the next two centuries.

Cook's achievements as a cartographer were appreciated most particularly by the Admiralty and the Royal Society. They commissioned him to undertake a voyage to the Pacific on His Majesty's Bark *Endeavour*, which lasted from 1768 to 1771 and would become known popularly, rightly or wrongly, as his "First Voyage," because almost always the additional words "in the Pacific" were left unsaid. Ostensibly, he was to make observations on the Transit of Venus, when the planet

Venus passes between the Earth and the sun, which it does at rare but predictable intervals. Differences in the paths of Venus when viewed from widely separated locations on Earth would be used to calculate the distance of the Earth from the sun. However, when Cook's observations of the transit from Tahiti were completed, he opened sealed orders from the Admiralty only to discover that he should head south to search for the mythical continent of *Terra Australis*.

In doing so, Cook sighted New Zealand on October 6, 1769, and by circumnavigating it, proved that it was not the postulated southern continent. He crossed Abel Tasman's sea and, while becoming the first person to map the eastern edge of Australia and showing that it was indeed of continental size, he also demonstrated that it was too far north to be the great lump of southern land that fellows of the Royal Society, principally led by the Scottish geographer Alexander Dalrymple, hypothesized was necessary to balance the Earth. Although, to be fair, the notion of *Terra Australis Incognita* had been proposed first by Aristotle and then by many others ever since.

The inconclusive results of his First Voyage only whetted the appetite of the public, the Admiralty, and the Royal Society for a second. Cook was promoted to commander and in 1772, aboard the HMS *Resolution*, he set sail on his Second Voyage to prove or disprove the existence of *Terra Australis* once and for all. This expedition would last until 1775 and it was this expedition that had been joined by Johann Reinhold Forster after Joseph Banks—who had, by then, become a bit of a prima donna and tried to take command of the expedition himself—pulled out.

While the Emperor penguin standing before me looking rather forlorn in the dark and uninspiring exhibit at Tring had not been collected by Forster, it seems likely that Forster was, indeed, the first person to ever see an Emperor penguin. Commander James Cook managed to take the *Resolution* farther south than any ship had ever been, crossing the Antarctic Circle on January 17, 1773, and continuing to explore southern waters, eventually reaching as far south as 71°10′S a bit over a year later. Cook had come within a piece of pack ice or two of discovering the Antarctic continent but, somehow, he managed to miss it. Nevertheless, the extreme southern latitude of one of Forster's supposed observations of a King penguin suggests that it was most likely an Antarctic-living Emperor penguin rather than the closely related but sub-Antarctic-living King penguin. An Emperor penguin looks sufficiently like a King penguin that it is completely excusable for even a naturalist of Forster's ilk that he might mistake it for one—especially from a distance and especially because, at the time, Emperor penguins were not known to exist.

It seemed from Cook's Second Voyage that *Terra Australis* did not exist either: he had managed to get seriously south and, apart from a few small islands that he claimed for king and country, the supposed southern continent was nowhere to be seen.

The lure to get farther south did not stop there, however. For the next 140 years, a good many men would endure extreme hardship, take unbelievable risks, and often lose their lives in an attempt to get farther south than any men before them. It was, indeed, a strange preoccupation and one that, a little over a century after Cook returned to England on the *Resolution*, would be acquired by a boy born on the banks of the River Tyne at the very place where Cook had started his seafaring career.

As I stand there before the faded but otherwise remarkably well-preserved body of a penguin that has been dead for 170 years, I realize that I know more about its life than I do about that of the boy from Newcastle. I know where it lived, when it bred, how it communicated, and what it ate. Yet, if I am to begin to understand Levick, to understand, for example, how and why he should have come to join those endeavors of men to reach ever southward, it seems like I have come to the right place.

Douglas Russell is everything I imagine someone bestowed with the title senior curator of birds' eggs and nests should be: he is tiny, bespectacled, and as twitchy as a sparrow picking up bread crumbs. He is wearing a waistcoat with jeans, but his beard and glasses make him seem more erudite than rebellious. He is also remarkably open and friendly.

Douglas leads me to his office through vast rooms containing what I imagine are the more than 300,000 clutches of birds' eggs and the more than 4,000 nests for which he is responsible, not to mention the nearly 750,000 bird skins the museum now contains (108 of the skins taken by Edwin Rist remain unrecovered). If Douglas were indeed a bird, he would be more bower bird than sparrow it seems: his own nest is overflowing with little treasures.

I ask him first about the unpublished manuscript of Levick's that he had discovered. It turns out that Douglas had gone to the library in the Rothschild building—for what, he cannot remember—but they did not have whatever it was and, to occupy his time, he began flicking idly through reprints held in filing boxes in the library. He happened to have the box before him with reprints of authors whose surnames start with L and, because he knew of Levick, he decided to look at Levick's papers. It was then that he spied the three-page manuscript with the words NOT FOR PUBLICATION printed across the top of its front page. Not realizing its significance at that stage, but intrigued by the notion that what it contained should be kept from the public, he decided to have a closer look.

It seems that the contents of the unpublished manuscript that Douglas had found were originally intended for inclusion in Levick's book *Antarctic Penguins: A Study of Their Social Habits*—at least, until the heavyweights at the British Museum of Natural History intervened.

Douglas conducted a thorough search through the records of the Natural History Museum and at first could not find any reference to the unpublished manuscript or any other surviving copies. That was, until he came across a note from Sidney Harmer, the quaintly titled keeper of zoology, written to William Ogilvie-Grant, the curator of birds. Harmer made it clear that the contents of Levick's manuscript about penguin sex should not be included in Levick's book from the expedition and that, instead, one hundred copies could be printed for internal consumption only.

Douglas looks at me with his habitually earnest stare, speaking in a strong, slow, matter-of-fact voice:

Harmer writes to Ogilvie-Grant and says we'll have it cut out,
we're not going to include it in the published version.

I ask Douglas if I can see the actual copy of Levick's lost manuscript
and he, obligingly, retreats to the vaults of the museum and returns
with a manila filing box. He removes the thin and, but for a couple
of dog-ears along its leading edge, remarkably pristine paper that had
caused all the fuss when he finally published it nearly a century after
it had been written. In fact, Douglas's sleuthing eventually did turn
up one other surviving copy of Levick's manuscript: it was tucked into
the private papers about the Terra Nova Expedition belonging to none
other than Walter Rothschild himself.

In 1977, when I first read Murray Levick's book about penguins, little did
I imagine that Levick had been gagged, that the book had been censored.
Or that, nearly forty years later, I would find myself sitting in an office
on the other side of the world, holding a manuscript written by Levick
describing the sexual behavior of Adelie penguins in such torrid terms,
and—as determined from my own research—in such accurate detail.

I put down the manuscript with its edges slightly darkened—perhaps
from the oil on Douglas's fingers, perhaps from that on those of Levick
himself—and stare back at Douglas equally hard. Except that I am not
really looking at him: I am reflecting that as I set about uncovering the
story about this man Murray Levick and penguin sex, it will, inevitably,
also involve me.

I ask Douglas if I can see Levick's original field notes—the notebook where Levick initially recorded all his observations of Adelie
penguins—which Douglas had referenced in his commentary published
in *Polar Biology*. He is hesitant. "No," he says. The field notes are held by
an antiquarian book collector who, Douglas explains, wishes to remain
anonymous. Yet, as we continue to talk in his office cluttered with bits
and pieces of birds, nests, paper, and God knows what else, a kind of

mutual respect and trust begins to develop between us. Douglas admits that he would like to have the field notes published. If Douglas can get permission to do so, it will be good to have a penguin expert to interpret Levick's observations. "Perhaps you might like to do that?" he asks.

I leave Tring with a promise from Douglas that he will contact the owner to see if we can view Levick's field notes together. He cautions me, however, that he cannot promise his request will be successful.

———

I am in a book-lined room on the third floor of an elegant apartment in one of the better suburbs of London. On an oval oak table at the center of the room, there is a blue-covered notebook. On its cover, in large awkward letters written with red paint, are the words ZOOLOGICAL NOTES CAPE ADARE VOL. I.

It is the book in which Murray Levick recorded his observations of the penguins at Cape Adare. Only four people, as far as I can discern, have had the opportunity to see these notes in their entirety during the more than one hundred years since they were written: Levick himself, the gentleman who owns this apartment and all the books it contains, my companion Douglas Russell, and me. I gently pull back the cover. The writing in blue-black fountain pen ink is taut, composed of small, neat letters. I randomly open the notebook at another page and stop. Stop everything. Stop talking, stop moving, stop breathing. It is just as Douglas had described: a section of the page has been pasted over with another piece of paper and other text that I cannot decipher. I can see that they are Greek letters but they make no sense to me. It's a code. For some reason Murray Levick, not just Sidney Harmer, had wanted to keep this part of his writing secret from prying eyes.

But why? These were zoological notes, were they not? Observations of penguins in Antarctica. Why the need for such secrecy; for such caution?

I glance out the window at clouds swollen with rain and I am transported back nearly two decades to another day with gray clouds. To the day when I saw two male penguins fucking each other.

A gust of wind rattles the window pane and I look down again at the blue notebook before me. I am reminded of that moment on January 16, 1912, when Captain Robert Falcon Scott saw Roald Amundsen's black flag and realized that he was not going to be the first to get to the South Pole. This notebook is indisputable proof that Murray Levick discovered the sexually depraved side of penguins long before I did. But—and this is the most curious thing of all—as this page with its pasted piece of paper and indecipherable Greek letters would seem to attest, he wanted to keep such sexual shenanigans largely to himself.

Why? That is what makes me hold my breath more than the physical evidence of Levick's observations or, even, his deception. It doesn't really matter who was the first to see a bit of male-on-male action in penguins any more than it probably matters who was first to stand on an arbitrary piece of ice and drive a flagpole into it. What could compel a man like Scott to sacrifice everything for the chance of doing so, however, is surely more interesting. And, it strikes me in that stilled moment, in that room with hundreds of antiquarian books but seemingly no oxygen, that uncovering why Murray Levick would discover the dirty side of penguins and then try to cover it up is surely more compelling than a bit of sodomy between consenting birds.

I turn the pages of the notebook over, photographing each one like it truly is evidence at a crime scene; like the detective I have now become.

I close the cover of the notebook. Outside, two men with billowing coats run one after the other but, from my third-story perch, I cannot make out if they are running to shelter from the impending rain or whether the second is chasing the first, intent on catching him and, at the very least, giving him a good hiding. I put my camera down beside the notebook. I feel a strange connection to the mysterious man who wrote it. If I were Scott, then Murray Levick would surely be my Amundsen. Yet we are no longer racing toward the same goal. Murray Levick is now my goal. I am chasing him. Not the South Pole. Not the murky behavior of penguins. Murray Levick.

CHAPTER THREE

THE THREE NORWEGIANS

C hasing Murray Levick is not as easy as simply running after him. I cannot see him ahead of me, and he has left so little in his wake. Even his destination is not clear. Levick may well have become the world's first penguin biologist, but, even at first glance, it is apparent that he never set out to study penguins. We share that in common. It was Antarctica, not penguins, that lured me, and I suspect Levick too.

When I was a small boy—pre-acne, pre-adolescence, pre-ambition—I became obsessed with Apsley Cherry-Garrard's book, *The Worst Journey in the World*. Cherry-Garrard, like Levick, had been a member of Scott's Terra Nova Expedition. His account of the expedition, a story of Antarctic heroism and adventure, gave me the ambition I had been lacking up till then: it established within me an absolute conviction that what I most wanted to do in life was go to Antarctica. It triggered a yearning for my own adventures and a desire to set foot on the Great White Continent that grew in me, year after year, like a benign but relentless cancer.

It is September 1, 1977, the first day of spring. I am a long-haired student at the University of Canterbury in Christchurch, New Zealand's gateway to Antarctica. The New Zealand Antarctic Research Program operates out of Christchurch and I have joined the university as a PhD student intent upon studying Weddell seals in Antarctica. While I have long had an interest in seal biology, it is Antarctica that really seduces me. I could have studied seals virtually anywhere but I have chosen to come to the University of Canterbury solely because research on the Antarctic-living Weddell seals is my ticket to Antarctica.

Christchurch is flat and staid to my twentysomething eyes. Yet, even the wet spring weather, which has replaced the fog and chill of winter, cannot dampen my spirits as my scheduled departure date approaches. In my dingy bedroom of the St Albans house that I call home, the world has never looked brighter. I pack, unpack, and repack. I can barely sleep; barely contain my excitement.

Then disaster. Followed soon after by utter devastation. Just six weeks out from my departure, I am told that the research project I have planned is no longer viable. What to do? I want to go to Antarctica so desperately that, if seals cannot be my means to get there, what can? That is when, for the first time in my life, a thought about studying penguins waddles into my brain.

It is October 18, 1977, and I am on a U.S. Navy Starlifter jet, dressed in yellow, puffy, down-filled survival clothes, surrounded by bearded and much older men. I am heading to McMurdo Station in Antarctica and, from there, I will take a helicopter to the Adelie penguin colony at Cape Bird on Ross Island.

After six hours flying, I stare out the small round window and I am awed by the rugged land below, which is covered completely by a sheet of white ice through which peeks a chain of mountains. Even from thirty thousand feet I can make out crevices in their glaciers. One enormous glacier, the Drygalski Ice Tongue, sticks far out into the blue waters of the Ross Sea, which is dotted with millions of brilliant white ice floes, looking for all the world like God's very own jigsaw puzzle as seen from the heavens. My pulse quickens with the anticipation of taking my first footsteps on the frozen continent below me, yet I know I am by no means the first to experience the lure of Antarctica.

⸺◈⸺

Given the vastness of Antarctica and the notion of *Terra Australis* existing for over two thousand years, it seems astonishing to me that the continent remained hidden and unseen until about two hundred years ago. The Russian explorer Fabian Gottlieb von Bellingshausen is credited with the first sighting of the Antarctic continent on January 28, 1820, with additional sightings by English and American sailors later in the same year.

Twenty-one years later and just four years into the reign of a young Queen Victoria, the English explorer James Clark Ross leads an expedition to Antarctica that discovers the Ross Sea, the large indentation in the continent of Antarctica sitting directly below New Zealand. The naturalist aboard Ross's ship the HMS *Erebus* is Joseph Dalton Hooker, who snatches some Emperor penguins from the sea ice. The expedition also consists of a second ship, the HMS *Terror*, captained by Ross's dear friend, Francis Crozier. Ross, Crozier, and the two ships leave their names on the features of Antarctica that will eventually become so familiar to both Levick and me.

The island sitting at the southernmost end of the Ross Sea is christened Ross Island, while its two volcanoes, one active and one dormant, are called Mount Erebus and Mount Terror, respectively. The easternmost point of the island, a breeding site for both Emperor and Adelie penguins, is named Cape Crozier. The northernmost point acquires the moniker, Cape Bird, after the first lieutenant on the *Erebus*, a not inappropriate appellation given that it is home to tens of thousands of black-and-white Adelie penguins each summer. It is destined to become my home too, the main base for my decades-long study of penguins.

In a surprising twist to the story, Ross, Crozier, the *Erebus*, and the *Terror* all leave their mark on the Arctic too.

Not long after Ross returns to England from the Antarctic in 1843, the ships *Erebus* and *Terror* are taken under the command of Sir John Franklin, with Crozier once more at the helm of the *Terror*, on an expedition charged with finding the much sought after, but hitherto unrealizable, Northwest Passage: a sea route connecting the northern Atlantic and Pacific Oceans through the Arctic Ocean. They sail from England in May 1845, and by September the following year, both ships have become trapped in the Arctic sea ice. Within a year, Franklin and many of his men are dead. Crozier takes over command of the remaining expedition members but, eventually, they too succumb. Without any word from the expedition for three years, the British Admiralty sends James Clark Ross to search for Franklin and his friend Crozier, which he does for the better part of two years to no avail. A reward of £20,000 leads to many more expeditions going in search of the Franklin Party: the grave sites of three of the crew are found on Beechey Island and, later, the remains of others on King William Island.

All 129 men of the Franklin Expedition perished. It remains, to this day, the largest single disaster and loss of life in the history of polar exploration.

And, more than a century and a half later, it remains in the news.

—

It is September 2, 2014. The icebreaker CCGS *Sir Wilfrid Laurier*, belonging to the Canadian Coast Guard, is in the calm waters of Queen Maud Gulf in the Canadian Arctic, surrounded by a sea and landscape christened with acknowledgements to Queen Victoria and the Franklin Expedition, which had gone missing on her watch: Victoria Island, Victoria Strait, and Franklin Strait. On board the icebreaker, two Parks Canada archaeologists are watching the output from a side sonar when they see the almost perfectly formed apparition of the hull of a ship lying in thirty-six feet of water like a ghost from the past, eerily well preserved and ethereal on the screen. No less than the Canadian prime minister himself, Stephen Harper, announces to the world that a

Canadian expedition tasked with finding Franklin's ships has, indeed, found the HMS *Erebus*. He describes it as, "a really important day."

I am in my office, sitting in my red chair reading the news on the BBC website. I stop my rocking, my hands grip the chrome arms of the chair so hard it is a wonder that I do not crush them. I am frozen before the brown, X-ray-like sonar images playing in the clip that accompanies the news report. The ship looks perfect save for its broken stern; the stern that housed the captain's cabin and might even now, I suppose, house Franklin's body.

In his piece to camera, the Canadian PM waxes lyrical about how exciting this all is and what it means to Canadian history and folklore—but none of that excites me. I am elated by what it all means for the southern extremities, not the northern extremities of the world. This is, after all, the ship on which Joseph Dalton Hooker collected the Emperor penguin that had stood to attention before me in the Rothschild Museum in Tring. This is the ship that gave its name to Mount Erebus, the most prominent landmark in the part of Antarctica I have come to think of as my second home. Most significantly, this is the ship whose final voyage would be responsible for eliciting the urge to explore polar regions in a young Roald Amundsen and, therefore, as I am beginning to realize, the ship that would set in motion the events that would lead to Murray Levick turning up at Cape Adare and undertaking his study of penguins. And here it is, staring back at me in all its sepia-colored glory, unexpectedly tangible and real, despite disappearing 166 years earlier.

It is beyond "important," to me, Stephen Harper.

That exhilaration is still with me as I find myself walking with quickening steps down a gravel path toward a white wooden house on the banks of the Glomma River in Norway. The sun is strong enough to have melted the recently fallen snow, though the air remains crisp and cold. As I approach the door, which is placed dead center on the front side of the house, the most telling object is not the bust of a man with a nose that would not look out of place on any bird of prey, but surely the pair of red wooden skis that are leaned against the wall on the left side of the door. They are very short and very old: the type of skis that might have been used by a young child near the end of the 19th century. For this is the place where on July 16, 1872, Roald Engelbregt Gravning Amundsen, who learned to ski almost before he could walk, was born four years before Murray Levick.

Roald Amundsen is the youngest of four sons of a Norwegian seafaring family. When he is two, the family moves to Christiania, the capital of Norway (changed to Kristiania in 1877 and now called Oslo). Their house there adjoins the forest, and Roald, his brothers, and a gang of other friends ski and play in the outdoors. One of his playmates, a boy eight years older than him, is Carsten Borchgrevink. Borchgrevink and Amundsen will prove to be crucial factors in Murray Levick ending up studying penguins. "Penguin biologist" was never an intended destination for Levick. As I am beginning to discover, he gets there through a series of unintentional circumstances: the Accidental Penguin Biologist.

Amundsen is eight or nine when the story of Sir John Franklin's lost expedition "captivated his imagination." It proved to be the pivotal point in his life, filling him with the desire and determination to become a Polar explorer, in much the same way that Cherry-Garrard's account of polar adventure and drama had so affected me as a young boy.

Amundsen's boyhood desires receive a huge boost in 1888 when his countryman, the Norwegian explorer Fridtjof Nansen, does what many before him have failed to do: he leads the first expedition to cross the Greenland ice cap. Nansen brings many innovations to Arctic exploration, including the use of light sledges, specially adapted clothes and tents, an efficient cooking system, and most particularly the use of skis rather than snowshoes for traveling across the polar regions. Nansen returns triumphantly to Norway in 1889 and, as he sails up the fjord to Kristiania, there among the welcoming crowds is a tall, somewhat short-sighted seventeen-year-old schoolboy.

As Roald Amundsen described it:

> *That day I wandered with throbbing pulses amid the bunting and the cheers, and all my boyhood's dreams reawoke to tempestuous life. For the first time something in my secret thoughts whispered clearly and tremulously: if you could make the North West Passage!*

And there it is. In that strange amalgam of childhood friend and childhood hero, an unlikely troika of three Norwegians emerges—Amundsen, Borchgrevink, and Nansen—whose lives will ultimately become entwined with that of a rugby-playing thirteen-year-old who is, at that moment, a thousand miles away in England. Through a series of unlikely events, they will conspire to lead him, admittedly accidentally, to embarking upon the world's first serious study of penguins. Yet, all this is still far beyond the ken of the young Murray Levick. His family had moved from Newcastle when he was just four. As a schoolboy at St. Paul's School in London, he showed an aptitude for sport—especially gymnastics and rugby—if nothing else. The lure of the polar regions and the hardships that lie ahead are still far away from his thoughts.

Not so for the seventeen-year-old Roald Amundsen. It is, perversely, the hardships of polar exploration after which he hungers most.

Yet, many obstacles stand in the way of Amundsen setting out on his ambition to be the first to complete the Northwest Passage, to try to do what neither Franklin nor Crozier could do in an area that even the great explorer James Clark Ross had found impenetrable. Not least, Amundsen's father dies when Roald is just fourteen, and his mother, upon whom he depends for financial support, is determined that he shall become a doctor. Amundsen duly enrolls at Kristiania University but proves to be as poor a student as he is poorly motivated for a life in medicine. Two events do, however, inspire him.

First, Nansen begins building a new ship for an audacious return to the Arctic, whereby he plans to allow the ship to get frozen in the pack ice and then be carried in the ocean currents across the Arctic, providing him with the opportunity to become the first person to reach the North Pole. His ship, the *Fram*, is revolutionary, with a rounded reinforced hull that is designed to rise up when squeezed by pack ice and, thereby, withstand the enormous pressures that would destroy other ships. In 1893, Amundsen is part of the crowds that once more line the shores of Kristiania in order to, this time, bid Nansen and the *Fram* farewell.

Second, that same year, Amundsen attends a lecture by Eivind Astrup, who had accompanied the American explorer Robert Peary on a 1,250-mile crossing of the Greenland ice cap on skis. Astrup is less than a year older than Amundsen. The previous year, when he was just twenty-one, Astrup had been awarded the Order of St. Olav, Norway's highest honor, making him the youngest ever recipient. Astrup is credited with pioneering the use of dogsleds combined with skiing for polar travel, using skills he learned from the Inuit. Amundsen is so inspired by his young countryman that he leaves the lecture and immediately, that night, sets out on a winter ski expedition in conditions that are suitably Arctic-like.

On September 9 of that same year, Amundsen's mother dies and, soon afterward, he leaves the university to pursue his own dreams rather than her dreams.

Today, the Fram Museum sits on the shores of Oslofjord in sight of where Amundsen stood to cheer on his childhood hero, Fridtjof Nansen. The sign on its glass door says that it opens at 10:00 A.M. I am fifteen minutes early, eager to get in. This is a museum devoted to polar exploration and, in pride of place, greeting me as I enter, is the *Fram* itself. It is unexpectedly large and looks in perfect condition, nearly as good, I imagine, as the day that Nansen commissioned it. The color palette of its rounded hull is striking, being painted with three equally wide swathes of white, black, and red. The ship itself is made of three layers of hardwood, with the bow reinforced by horizontal strips of iron. It looks as hard as granite; as solid as any rock.

I lay my hand upon its hull and feel a spasm of electricity—a direct connection with Levick that is not dissimilar to the one I had experienced when first leafing through the pages of his Zoological Notes in a London flat. I set about availing myself of the museum's resources to help decipher why this connection to the world's first penguin biologist through a Norwegian ship should feel so powerful to me.

The place to start, I decide, is with another Norwegian ship and another Norwegian polar explorer: Amundsen's childhood friend, Carsten Borchgrevink.

Eleven days after the death of Amundsen's mother, a ship called, appropriately enough, *Antarctic*, sets out from Norway on an expedition organized by Norwegian entrepreneur Henrik Bull to go sealing and whaling in subantarctic and Antarctic waters. When the *Antarctic* passes through Melbourne on its way south, Carsten Borchgrevink, who had earlier emigrated to Australia, convinces Bull to take him on as a deckhand and part-time naturalist. Unable to find many whales around the subantarctic islands, Bull and the ship's captain, Leonard Kristensen, decide to take the ship south into the Ross Sea. There, on January 24, 1895, they lower a small boat, and a party of seven men, including Bull, Kristensen, and Borchgrevink, make the first recorded landing by men on the Antarctic continent. The place, a spit of land on the northern extremities of the Victoria Land coast, had been named half a century earlier by James Clark Ross as Cape Adare. By chance, this spit of land just happens to contain the world's largest breeding congregation of Adelie penguins.

From the outset, then, the first tentative footsteps of humans on the Antarctic continent have become crossed with those of the Adelie penguins. They will, through the cascade of events I am trying to uncover, lead to Murray Levick sitting among the penguins seventeen years later, writing observations in a blue-bound notebook. Given that some Adelie penguins can live up to twenty years or more, and given that the Cape Adare colony contains well over half a million birds, some of the penguins that watch those first humans arrive in their midst must also be there seventeen years later when Levick, eventually, takes his own tentative footsteps on the same spit of land.

In between, the penguins must endure a return visit by one of those men who rowed ashore in the small boat near the end of their breeding season in January 1895.

The first of the boat's occupants to actually set foot on the Antarctic continent is either Borchgrevink or Kristensen. Both claim to be first. However, Borchgrevink departs the *Antarctic* in Melbourne as it sails home and, by the time the ship reaches Norway and Kristensen can make his claim of priority, Carsten Borchgrevink has already told the world that he was first and it will be Borchgrevink who evermore is known as the first man to set foot on the Antarctic continent. Whether a giant lie or a giant leap for mankind, it is, either way, a small yet important step on the road to uncovering the sexual antics of Adelie penguins.

Borchgrevink thereafter rushes to the Royal Geographical Society in London where, on August 1, 1895, at the Sixth International Geographical Congress, he proposes mounting a scientific expedition to Cape Adare. He describes Cape Adare in benign terms:

> ". . . here the unbound forces of the Antarctic Circle do not display the whole severity of their powers."

If not a deliberate lie, Borchgrevink cannot have been further from the truth. However, while the congress recognizes the importance of Antarctic exploration, the president of the Royal Geographical Society, Sir Clements Markham, harbors plans for the society's own expedition to Antarctica and he fiercely opposes Borchgrevink's proposal.

Without Markham's support, Borchgrevink goes in search of private funding for his proposed expedition to Cape Adare. Eventually, much to Markham's chagrin, he finds it in the form of British publisher Sir George Newnes.

As it turns out, it is not Markham that Borchgrevink need worry about. Other competition is coming from the most unlikely of quarters.

———

Meanwhile, just as Borchgrevink has been stepping onto the Antarctic, Nansen and the *Fram* have been drifting in the Arctic. After leaving Norway, Nansen took his ship and embedded it in the dense pack ice north of Siberia, where, as predicted, it drifted with the currents that were moving the ice in what was mostly a northerly direction. However, the speed of their drift is much slower than Nansen has anticipated so he hatches a Plan B. When the *Fram* reaches 84°N, he and Hjalmar Johansen abandon the ship, leaving the remaining crew to continue their drift across the Arctic, while Nansen and Johansen take a dogsled and make a dash for the North Pole. Even with the dogs, progress is slow and when they get to 86°13′6″N, the farthest north any human has ever been, Nansen decides they must turn back, leaving the prize of the pole unclaimed.

After an extraordinary seventeen-month-long journey full of danger and daring, using initially the dogsled and then kayaks that they had built while on the *Fram*, the pair are eventually reunited with the *Fram* in the Norwegian city of Tromsø. Altogether, the *Fram* had spent three years drifting in the Arctic ice. Following their long years of absence, on September 9, 1896, Nansen and the *Fram* return triumphantly to Kristiania, confounding the experts who had predicted their demise. The *Fram* and every single one of the men on the expedition has survived.

The same cannot be said for all polar explorers. In January of the same year, the body of the twenty-four-year-old Eivind Astrup is found in the snow near the Norwegian village of Hjerkinn: he has shot himself with a revolver. The blood-soaked snow surrounding his body is a bleak reminder that one of the biggest dangers to polar explorers stems not from the cold or starvation but, rather, their own psychology. What depressed Astrup sufficiently to take his own life is unknown, but just as suicide is destined to become an all too frequent occupational hazard of polar explorers, so too, it turns out, are their sexual infidelities. And it is likely that the one may feed off the other. While Fridtjof Nansen has been away trying to reach the North Pole using the dogsled and ski method devised by Astrup, it is rumored that Eivind Astrup has been having an affair with Nansen's wife, Eva. Sexual misdemeanors in the polar regions are not, it would seem, the province of Adelie penguins alone.

———

If the North Pole is still up for grabs, so too are other prizes. Inspired by Borchgrevink, a Belgian naval officer, Adrien de Gerlache, decides to mount his own expedition to Cape Adare, where he intends that he, not Borchgrevink, shall become the first man to winter on the Antarctic continent.

Earlier, de Gerlache had been bitten by a similar bug to that which had so infected the young Amundsen and would eventually infect me: he wants desperately to go to Antarctica. Perhaps not surprisingly then, in 1896, as de Gerlache readies his ship the *Belgica* for the journey south, who but Amundsen himself should volunteer to serve on the expedition without pay. Amundsen has by then obtained his Mate's Certificate through working on sealing ships off the Greenland coastline in order to get polar experience. He is accepted by de Gerlache as second mate.

The *Belgica* leaves for the Antarctic in 1897. From Amundsen's perspective, the best thing about the expedition is its American doctor, Frederick Cook, who had explored Greenland with Robert Peary. Amundsen sets about gleaning everything that he can about polar travel from the Peary-trained Cook.

Otherwise, the expedition is mostly a disaster, teetering at times on the brink of their annihilation. The pack ice proves too much of an obstacle for de Gerlache to get to Cape Adare, so he decides they should become the first men to spend a winter in the Antarctic by emulating

Nansen and driving the *Belgica* into the midst of the pack ice until the ship becomes entrapped. Unlike Nansen, de Gerlache is ill-prepared for the cold, the darkness, and, in particular, the ravages of scurvy from a diet deficient in vitamin C. Many of the men, de Gerlache included, suffer from scurvy, and in the brutal cold and dark isolation become physical and psychological wrecks. Cook saves them all by insisting that they eat seals and penguins, a rich source of vitamin C. Well, all but one: a crew member refuses to eat the seal meat and dies.

It is not just vitamin C that they lack. One of the unspoken aspects about polar exploration is the sexual deprivations it imposes upon those undertaking such missions, which are often measured in years. Arguably, those in the north might fare better than those in the south given the presence of the local Inuit. Robert Peary is with his wife only three years out of their first twenty-three years of marriage, but in addition to the two children he has with his wife, he also has an Inuit mistress and fathers at least two children with her. But there is no such salve for loneliness in the South. Yet the diaries of the men going there are characteristically silent on the subject of sex or, even, their desires for sex. Men of the Victorian era are not expected to voice such thoughts, even though, doubtless, they must have them.

The *Belgica* is a notable exception because Lieutenant Georges Lecointe, the second in command, produces a publication called, suggestively enough, *The Ladyless South*. In it, he ascribes to Amundsen the comment, "Yes, sir, I love it," when referring to the absence of women. Through high school, university, and afterward, Amundsen had displayed little interest in the opposite sex. He exhibits an aura of monasticism mixed with a spoonful or two of misogyny, all of which ensure that he is happiest when he is on ice. He professes no need for the warmth of a woman. If Victorian values suppressed sex, or at least the overt demonstration of it, it would seem that such social mores are wasted on Amundsen: apparently, he has no need for sex anyway.

When the sun returns on July 23, 1898, the problems for the *Belgica* and those on board are far from over. They sit imprisoned in the pack ice with no foreseeable way out.

As the *Belgica* sits locked in the sea ice of Antarctica's Graham Land, thwarted from reaching its destination of Cape Adare, Carsten Borchgrevink is setting out from England on his ship the *Southern Cross* on the somewhat inappropriately named British Antarctic Expedition: the money may have come from Sir George Newnes, but twenty-six of twenty-nine in the crew are Borchgrevink's Norwegian countrymen.

It takes the *Belgica* seven months to finally get free of the ice. Once more, it is Frederick Cook who saves them: he proposes cutting and blasting a channel in the ice to reach an open lead of water about a mile away. On February 15, 1899, the channel opens sufficiently for the *Belgica* to finally start her engines and begin the tortuous journey home. By then, Borchgrevink is just two days sailing from Cape Adare.

All in all, the Belgica Expedition is a lesson to Amundsen about the importance of being prepared for conditions in the high latitudes. Perhaps most significantly of all, it gives Amundsen the excuse to contact Fridtjof Nansen after getting back to Norway.

I stood in Nansen's villa at Liysaker and knocked on the door of his study. "Come in," said a voice from inside. And then I stood face to face with the man who for years had loomed before me as something almost superhuman: the man who had achieved exploits which stirred every fibre of my being.

The friendship forged between these two Norwegians in that moment is destined to impact Murray Levick. Yet, in 1899, while Amundsen, Nansen, and Borchgrevink are setting in train the events that will drive Levick to become the world's first penguin biologist, Levick himself is proving himself more capable than Amundsen in at least one regard. He is studying medicine at St Bartholomew's Hospital in London, and although he does not distinguish himself as a great student, he is at least sticking with his studies. There are no indications, however, that thoughts of Antarctica have ever crossed his mind and, certainly, no penguins.

Ironically, while Levick has left behind so little evidence of himself, the two ships that were the first to penetrate the Ross Sea—arguably the initial step in the train of events leading Murray Levick to Cape Adare—continue to reveal themselves despite nearly 170 years passing since they were lost and sunk during Franklin's expedition.

⁕

It is September 3, 2016. History repeats. A Canadian expedition finds the HMS *Terror* sitting in eighty feet of water in Victoria Strait, some sixty miles due north of where the *Erebus* had been found almost exactly two years earlier. Appropriately enough, it is found in Terror Bay off the coast of King William Island. Apart from a coating of kelp and other marine life, it is similarly well preserved.

Notes left by the crew reveal the *Erebus* and *Terror* became locked in the ice on September 12, 1846, and two years later those still surviving out of the original 129-member crew abandoned the ships. Yet none lived long enough to be found and rescued. It is a fate that so easily could have befallen de Gerlache and the men of the *Belgica* were it not for Frederick Cook's ingenuity.

Carsten Borchgrevink, as he crashes through the pack ice on the *Southern Cross* during his approach to Cape Adare, is much better prepared than either Franklin or de Gerlache, but he is not prepared for Cape Adare itself. It has been the lure of Antarctica, the mystique of the unknown frozen continent, which had taken him to Cape Adare in the first place and is bringing him back again now. It is the same lure that will eventually bring men like Scott and Levick there and, ultimately, me. Yet Borchgrevink and his men will come to loathe the place.

Some lure!

PART TWO

ALL ROADS LEAD TO CAPE ADARE

Divorce

We are told, overtly and covertly, that monogamy is our natural state. Marriage ceremonies and the Christian Church don't just extol monogamy as a virtue, they demand it "till death do us part." And, to a certain extent, we scientists have contributed to this view too. For sure, male red deer and bull elephant seals are obvious exceptions where polygamy reigns. But in those creatures where secondary sexual characteristics like antlers, enormous size, or red breasts are absent, leaving males and females essentially looking alike (monomorphic in the *lingua franca* of the scientist), scientists argue that mating with one long-term partner is not a marriage of convenience so much as, biologically, the best policy. This is because in animals where the sexes look alike, they are often monogamous and, when a pair breed together, like so many things, "practice" really does "make perfect": experienced pairs have a higher likelihood of breeding success compared to newly formed pairs.

If you believe the cartoonists, no animals look more alike than do penguins. And there is an element of truth to that: the evidence suggests that visual clues alone are not enough for male penguins to distinguish females, let alone individuals. On average, males are, typically, somewhat bigger than females but there is so much overlap between the sexes that it is not a reliable guide to who likes being on top. Might this partially account for Levick's and my observed instances of homosexual mountings: mistaken identity?

There is apparently no strong evolutionary advantage for characteristics like large size or ostentatious plumage in male penguins that might allow them to compete for females. That is because nearly all males, irrespective of whether they are the penguin equivalent of Don Juan or Quasimodo, get to breed and get to breed only once.

This is a consequence of where penguins feed (the sea) compared to where they breed (the land). No female penguin could manage alone if she were required to lay the eggs, incubate the eggs, and then simultaneously brood the chicks while also getting food for them and herself. It is a problem peculiar not just to penguins but nearly all seabirds, and it is the reason why virtually all seabirds are monogamous and, therefore, monomorphic. A male penguin cannot afford to leave his partner to pursue other females if he is to be successful in an evolutionary sense: he must remain with his partner and help her to raise their offspring.

Furthermore, the longer a pair stays together, the greater is the likelihood that their prior experience of breeding together will better enable them to fledge their chicks successfully. Consequently, even the scientists joined in the chant that penguins mate for life.

But, when all the participants in a soap opera look identical, it is hard to see the opera, let alone the soap.

CHAPTER FOUR

FIRST OBSERVATIONS

I t is February 17, 1899. The small whaling ship *Southern Cross*, after battling dense pack ice and storms for over seven weeks, makes its way into the unexpectedly open and ice-free waters of Robertson Bay at Cape Adare. The crew finds itself in an amphitheater formed by dark, stupendously steep mountains. Gravity-defying glaciers cling to their sides. Gigantic crevices, big enough to easily swallow the entire ship, are clearly visible from its deck. Beyond the mountain tops lies the interior of Antarctica and the wind that bears down on them comes from there, cold and fierce. It feels more prison than sanctuary. There is just one place where they can manage a landing: a flat spit of land that is all but deserted of the penguins that had covered it when Carsten Borchgrevink first set foot there five years earlier. Borchgrevink's men are impressed, but not in a good way:

> It seemed, at a distance, so small and inhospitable that some of my staff felt constrained to remark at first sight of the place, that if it was there I proposed to live for a year, they had better send letters of farewell back with the vessel.

They use whale boats to ferry equipment, food, and seventy dogs ashore. They take two weeks to establish their camp which they named "Ridley Beach," during which time they are belted by gales and pelted by rocks. At times they are unable to stand in the near-constant wind. A prefabricated wooden hut of Norwegian design, dark and solid, is erected upon the remains of penguin nests. Then, on March 2, 1899, the ship sails away to spend the winter in New Zealand, while Borchgrevink and his nine men are left alone: the most isolated men in the world, vying to become the first humans to winter on the Antarctic continent; that is, if they can survive that long.

All does not go quite to plan. It is mercilessly cold and windy, not the benign place that Borchgrevink had supposed from his earlier visit. There is little structure or discipline and the men become moody and morose. During winter, the sea freezes in Robertson Bay and around Cape Adare. Borchgrevink, bravely, undertakes several weeks of sledging, exploring the surrounding area during the heart of the Antarctic winter, when he and his two Finnish Laplander companions, Per Savio and Ole Must, are without sunlight and in temperatures that drop as low as -52°F. The cold is exaggerated greatly by the incessant wind. After the winter, however, the ice in Robertson Bay breaks out, and the men find the opportunities for sledging and exploring are limited by the surrounding mountains. There is nowhere they can go. They are trapped.

The expedition has a zoologist, Nicolai Hanson, who had been a contemporary of Roald Amundsen at Kristiania University, although he has the distinction of having actually graduated. Hanson is a handsome man with thick, wavy dark hair, a rock for a chin, and a thick mustache that, in the Antarctic, he wears in combination with a thick goatee. He married not long before leaving for Antarctica and left behind a pregnant wife. His daughter, Johanne, was born just over a month after his departure.

Given their enforced internment in the midst of the penguin colony on the spit at Cape Adare, which Borchgrevink has named Ridley Camp, it is certain that Hanson would have set about conducting the world's first study of penguins. But tragedy strikes. Hanson had become very ill on the five-month journey from England to Cape Adare. Although he recovers enough to begin his zoological work when at Cape Adare, toward the end of the brutal winter, his intestinal problems return and his condition deteriorates, leaving him bedridden.

On October 14, 1899, the first Adelie penguin is spotted arriving at the Cape Adare colony to begin the new breeding season. Hanson asks to see it, and one of the men obligingly catches it and brings it to his bedside. Hanson is truly excited to see the penguin. It is the last observation he will make. Thirty minutes later, he dies.

Rather than becoming the world's first penguin biologist, Hanson, instead, gains the far more dubious distinction of becoming the first person to be buried in Antarctica. Even that proves almost out of his reach. The Australian physicist Louis Bernacchi and the two Finnish Laplanders are sent to dig the grave in the place that Hanson has picked out for himself: on the north side of a large boulder one thousand feet up on the point of Cape Adare. The ground is frozen and almost impossible to dig so that, after working all day and breaking all their implements, they have, as Bernacchi puts it, "only succeeded in excavating to a depth of about 4 inches." He adds wryly, "On the next day we brought up a large quantity of dynamite, and, by its aid, we were more successful."

In exchange for the opportunity to see Adelie penguins, Hanson has lost the opportunity to ever see his daughter. The cost to his daughter proves equally high. Johanne Hanson Vogt will die in 1999, exactly one hundred years after her father is buried in the loneliest grave on Earth; an extraordinarily long time to live without a father.

When the *Southern Cross* finally arrives to pick up the nine surviving members of the party at the end of January 1900, the last thing they do is visit Hanson's grave and erect a small black cross with a brass plate bearing his name. Bernacchi sums up their departure:

> *It was one of the most bleak and ungenial days imaginable . . .*
> *We were not sorry to leave that gelid desolate spot, our place of*
> *abode for so many dreary months.*

Borchgrevink is no zoologist but, even without Hanson, he is intent on having the expedition recognized for its scientific achievements. At heart, however, Borchgrevink remains the same man who claimed to have put the first boot on the Antarctic continent: he is a man of firsts. Now back on the *Southern Cross*, he takes the ship south into the sea that has not been traveled since James Clark Ross discovered it in 1841 and lent it his name. Borchgrevink makes a number of "first landings" at places that would later figure large in the story of Antarctic exploration and penguin research: Mount Melbourne, Cape Washington, and Franklin Island.

Franklin Island epitomizes the strange but profound interconnectedness between those trying to explore the opposite ends of the Earth. It had been named by Ross in honor of Sir John Franklin, the Arctic explorer. The very Franklin who four years later would take Ross's ships, *Erebus* and *Terror* (along with Ross's friend Crozier), and attempt to pioneer the sought-after Northwest Passage. The same Franklin for whom Ross would search for two years. The same Franklin who would inspire Roald Amundsen to become a polar explorer, and whose childhood friend would then be the first to set foot on this Antarctic island named in Franklin's honor.

On February 10, 1900, Carsten Borchgrevink adds another milestone to his collection of firsts: he becomes the first person to set foot on Ross Island. It is very nearly his last. Borchgrevink and Bernhard Jensen, the captain of the *Southern Cross*, land at a beach between Capes Crozier and Bird on a small headland that Borchgrevink names Cape Tennyson. Soon afterward, a large iceberg calves off from the nearby glacier and "some thousands of tons of ice fell into the sea with a terrific and reverberating roar," creating a twenty-foot-high tidal wave. The two men are smashed against the rocky cliff at the back of the narrow beach, thumped by pieces of ice in their backs and completely swamped by the surge of water. Somehow, they manage to cling to the rocks and survival.

Thereafter, Jensen and Borchgrevink take the *Southern Cross* southeastward, noting the largely ice-free and gently sloping area that forms the Cape Crozier Adelie penguin colony. They follow the Ross Ice Shelf eastward, a flat and unbroken 160-foot-high sheer-walled barrier along which they sail for days until, on February 16, they make a discovery: an indentation in the shelf forming a natural harbor where a ship can be moored to the ice and access gained to the shelf itself. Borchgrevink lands on the ice shelf, another first, and with two men travels ten miles south, reaching 78°50′S, "the farthest south ever reached by man" as he will boast proudly afterward.

It is Borchgrevink's discovery of this indentation in the Ross Ice Shelf, I realize, even more than his landing and wintering over at Cape Adare, that will become the pivotal factor in taking Murray Levick on an unexpected journey that will end with him becoming the world's first penguin biologist.

When the Southern Cross Expedition finally arrives back in England, their reception is tepid. The country has its attention focused on a different Antarctic expedition: the forthcoming Discovery Expedition to be led by the Englishman, Robert Falcon Scott.

Borchgrevink is under an obligation to publish accounts of his expedition in George Newnes's publications, with a direction that he should make them accessible to the readers, and consequently less scientific in tone. Arguably, a laudable goal, it is nevertheless looked down upon by Sir Clements Markham and the other fellows at the Royal Geographical Society.

In 1901, Borchgrevink publishes his book, titled predictably, if not appropriately: *First on the Antarctic Continent*, in which he sets down his observations, including a chapter devoted to living among the Adelie penguins that were his neighbors on Ridley Beach. However, his observations provide little insight into the penguins that a trained zoologist like Hanson might have brought.

> *We all watched the life of the penguins with the utmost interest,*
> *and I believe and hope that some of us learnt something from*
> *their habits and characteristics . . . From the 14th October one*
> *continual stream of penguins waddled over the ice towards their*

summer residence; like so many people, they walked after one another. . . During the time of love-making—when they were studied most attentively by my bachelor staff—it goes without saying that they had many hard fights.

Borchgrevink manages to set the tone for all the observations of penguin behavior to come after his: anthropomorphic in style, with an analogy to humans, albeit the whiff of a notion that penguins are somehow better than us. These black-and-white upright little creatures are portrayed as paragons of a type of virtue that we should seek to emulate and, never more so than when it comes to their lovemaking.

However, Borchgrevink and his men had been much more interested in eating the penguins than in studying them. They rarely met a penguin on the journey south to Cape Adare that they did not kill. Throughout their summer spent at Cape Adare, they ate roasted penguin on many of the days of the week. When the penguins' eggs were laid, they had eaten those, with relish too. Borchgrevink takes a kind of pride in his book when he notes that by November 15 they had collected four thousand penguin eggs and secured them in salt as reserve rations.

Hanson was evidently not the only casualty of the Southern Cross Expedition. However, Cape Adare and its penguins were destined to have other expeditions inflicted upon them.

It is January 9, 1902. The HMS *Discovery*, under the command of Captain Robert Falcon Scott, pushes through the pack ice at the head of Robertson Bay and makes landfall for the first time in Antarctica at none other place than Ridley Beach, that tiny spit of shingle beach at Cape Adare that is home to a vast colony of Adelie penguins; the very same place that the first footstep had been placed on the Antarctic continent; the very same place that Borchgrevink had left behind just two years earlier after his party had become the first to overwinter on the frozen continent. In fact, among the crew onboard the *Discovery* is Louis Bernacchi, the physicist who had been one of those with Borchgrevink during those long dark winter months.

Robert Falcon Scott had been born on June 6, 1868, near the naval base of Devonport on the south coast of England. He is four years younger than Borchgrevink, four years older than Amundsen, and just over eight years older than Murray Levick. Given the military tradition in his family, it was preordained that he should join the Royal Navy. When just thirteen, he began his naval career as a cadet. By the age of fifteen, he was serving on ships in South Africa and, later, was stationed in the West Indies. It was there, that as an eighteen-year-old midshipman, Scott first caught the eye of the Royal Geographical Society's Clements Markham, who was impressed by Scott's abilities and attitude. In early June 1899, by then just turned thirty-one, Scott chanced upon Markham in a London street while home on leave. He learned, for the first time, of Markham's plan to send a ship, the *Discovery*, on a Royal Geographical Society expedition to the Ross Sea region of Antarctica. A few days later, Scott turned up at Markham's house to volunteer to lead the expedition to Antarctica.

Scott had been given command of the Discovery Expedition finally to realize Sir Clements Markham's vision of a British-led expedition to Antarctica. They had set sail from England in August 1901 with instructions to explore the Ross Sea and to get as far south as possible, perhaps even to the South Pole.

On board the *Discovery* with Scott are two others destined to become household names in the annals of what would become known as the Heroic Age of Antarctic exploration: junior doctor Edward Wilson and Third Officer Ernest Shackleton.

It is a time that marked the beginning of a new age in morality as well. In January 1901, after sixty-four years on the throne, Queen Victoria had died. She has been succeeded by her son, King Edward VII, who is as famous for his litany of affairs with married women as Victoria has been for her prudish ways—to the extent that he is otherwise known as Edward the Caresser. Outwardly, morally, the times are indeed a-changing. When the *Discovery* left England, it had been seen off by the country's new, sexually indiscrete king.

The expedition did not begin well. They traveled south via New Zealand and, as they bade farewell to the crowd that had gathered around Lyttelton Harbor to see them off, seaman Charles Bonner fell to his death from the top of the mainmast. He was buried two days later in Port Chalmers, when the *Discovery* made a brief stop in the seaport where I now live.

———

I had never been to the cemetery in Port Chalmers, and I am surprised by how picturesque it is, situated on a ridge overlooking the harbor out of which the *Discovery* had sailed, leaving behind the luckless Bonner. His grave is easily discerned. It sits near the top end of a line of flat and simple graves that stretch down toward the sea. Marking his grave, however, there is a large white marble spire that is taller than me. A plaque records that he died on December 21, 1901, aged twenty-three. I am stilled by that and brought to my knees. I was the same age when I boarded a Starlifter bound for Antarctica for the first time. I feel terribly for him. I know exactly the sense of excitement and youthful exuberance he must have been feeling when he climbed the mast to wave goodbye on his way to Antarctica. He deserved more than just a fancy headstone.

Charles Bonner's grave is an eternal reminder of the pitfalls of those times. So much of the Heroic Age would turn out to be accident, near accident, or accident waiting to happen. The heroic part actually came about in the ways that the men—and they were all men—dug themselves out of holes of their own making.

———

At Cape Adare, it is the penguins and not the ever-present dangers that seem to most impress Scott. He is fascinated as to why they are not content with just nesting on the flat ground of Ridley Beach. Instead, he observes, many of their number form "their nests on the steep hillsides, even to a height of 1,000 feet." He is equally impressed with the first human nest, noting that Borchgrevink's hut "is in very good condition," though he reveals something of his sentimental makeup when he remarks that:

> There is always something sad in contemplating the deserted dwellings of mankind, under whatever conditions the inhabitants may have left.

Certainly, Cape Adare is not appealing to Scott or his men as a prospective home, even though Robertson Bay is where Sir Clements Markham had suggested Scott should set up his winter quarters. Apart from its distance from the pole, there is the presence of the penguins. Edward Wilson, the expedition's junior doctor and zoologist, describes his reaction to the place in a way that suggests that he, like Hanson, will forgo the opportunities presented to him to become the world's first penguin biologist:

> There were literally millions of them. It simply stunk like hell, and the noise was deafening.

The next day, the *Discovery* pulls up anchor and heads south.

At first Scott follows in Borchgrevink's wake. He takes the Discovery ever eastward along the face of the Ross Ice Shelf until they sight new undiscovered land that forms the eastern edge of the shelf. He names this King Edward VII Land in honor of Britain's new king.

Turning back toward the west, in an area close to the inlet where Borchgrevink discovered that it was possible to land upon the shelf, Scott takes the ship into a small bight. Here the top of the shelf slopes down to the height of the ship. They moor next to the ice edge and are able to simply step off onto the ice.

It is then that Scott demonstrates the two elements that define his curious chemistry, which are capable of producing a reaction that results in either greatness or disaster, with the outcome depending largely upon luck: his brave determination coupled with a wanton disregard for adequate preparations. It was Joseph Dalton Hooker, the naturalist with James Clark Ross who collected my friend the Emperor penguin that is on display at the Rothschild Museum, who had first suggested that using a balloon in Antarctica would allow a visual assessment of its interior. Scott has brought with him two balloons and a number of hydrogen cylinders. What he has not brought with him, however, is anyone with much experience at ballooning. Furthermore, on February 4, 1902, he takes it upon himself, having never been ballooning before, to make the first flight in Antarctica in the one-man balloon.

Even as Scott himself tells the story, it is not a very smart move:

> *The honour of being the first aeronaut to make an ascent in the Antarctic regions, perhaps somewhat selfishly, I chose for myself, and I may further confess that in so doing I was contemplating the first ascent I had made in any region, and as I swayed about in what appeared a very inadequate basket and gazed down on the rapidly diminishing figures below, I felt some doubt as to whether I had been wise in my choice.*

At one point, Scott throws all the ballast out at once instead of jettisoning it gradually. The balloon shoots up in the air, oscillating wildly. Wilson manages to sum it up best: "The Captain, knowing nothing whatever about the business, insisted on going up first and through no fault of his own came back safely."

Scott had risen to a height of just under eight hundred feet. He is followed in the balloon by Shackleton, who takes a camera with him and gets even higher. And in that single act, Shackleton, too, demonstrates two facets of his own chemistry: his competitive relationship with Scott and his focus on gathering as much data as he can to support exploration in this unknown region of the world that is ladled with dangers and surprises.

The balloon flights and Shackleton's photos reveal that the Ross Ice Shelf is a flat plain of ice stretching as far as the eye can see to the south, a veritable highway to the South Pole. Scott names the inlet in the ice shelf Balloon Bight, but really it is part of the same geological phenomenon that formed Borchgrevink's earlier discovery of the nearby inlet. In this area, instead of floating on a watery base, the Ross Ice Shelf is being shunted over an island, which deforms it and produces these indented sloping edges of the shelf that can provide easy access to the interior.

His first and only ballooning experience behind him, Scott sends the *Discovery* westward, eventually setting up base in the most southerly reaches of McMurdo Sound. The *Discovery* is secured to the shore, allowing the men to use the ship as their winter base. They erect a hut for storing supplies next to the ship on what will become known as Hut Point. Given the predilection of Antarctic explorers to find themselves in tough situations, over the years, the hut at Hut Point will function as a refuge for many including Wilson, Shackleton, and Levick. Scott's decision to moor the ship there, however, comes with consequences: it gets completely frozen in over winter and cannot be freed the following summer.

In 1900, the year before leaving England for Antarctica, Scott had gone to Kristiania to consult with none other than the tall, blond, handsome Norwegian Fridtjof Nansen. Scott was keenly aware of the lack of British expertise when it came to sledging. He tried to pick Nansen's brains, to glean what he could, but Scott, pressed for time, did not spend long with Nansen. He bought some equipment and provisions from the Norwegians but opted to produce the rest in England.

Although Nansen had recommended the use of dogs for pulling sledges, Scott has brought with him only two men who have any experience with managing dogs and sledges. It is like the ballooning debacle all over again. Neither can the British men of the Discovery Expedition ski. Their first attempts to ski on the sea ice around the ship when it is moored at Hut Point are discouragingly bad, if one is Scott, or ludicrous, if one is Nansen, Borchgrevink, Amundsen, or any other Norwegian.

Despite the inadequacies of their preparation, Scott, Wilson, and Shackleton attempt the improbable: a trek to the South Pole with skis and dogs to pull sledges. Eivind Astrup, had he not shot himself, would no doubt have approved of their method of travel, and he himself may well have triumphed. As it is, while Scott and his men manage to get farther south than anyone before them, poor preparations and scurvy eventually take their toll and almost their lives. They have to retreat when still five hundred miles from the Pole. On the way back, Shackleton collapses and has to be helped by the other two. Once back, Scott orders Shackleton to be invalided home to England aboard the relief ship, the SY *Morning*, which has brought down more supplies to the men on the *Discovery*, which is locked in the ice.

Scott's subsequent account of their southern journey, which covered over nine hundred miles in three months, will anger Shackleton: it being implied that it was Shackleton's breakdown that led to them faltering when so far from the Pole.

Certainly, the dismal failure of their attempt to reach the South Pole is enough to convince Scott that the best way to travel in the Antarctic is to man-haul sledges without dogs. While he admits that dogs may potentially be faster if one is prepared to kill the dogs to feed those that are left, the experience of doing so on their own journey, when it became necessary to sacrifice all their dogs in order just to survive the return to their ship at Hut Point, leaves Scott averse to ever using dogs again:

> . . . *it left in each one of our small party an unconquerable aversion to the employment of dogs in this ruthless fashion . . . the introduction of such sordid necessity must and does rob sledge-travelling of much of its glory. In my mind no journey ever made with dogs can approach the height of that fine conception which is realised when a party of men go forth to face hardships, dangers, and difficulties with their own unaided efforts, and by days and weeks of hard physical labour succeed in solving some problem of the great unknown. Surely in this case the conquest is more nobly and splendidly won.*

And there, in his own words, Scott unwittingly sets out the reasons for his later failures. Furthermore, it is the same Victorian moral code that gives English gentlemen of his ilk an aversion to killing dogs, even when their own survival is at stake, that also prevents their talking about sex.

CHAPTER FIVE

BOYHOOD DREAMS

M urray Levick is a long way from even seeing penguins have sex, let alone being able to talk about it, but, unbeknownst to him, he is taking the first steps toward that end.

It is 1902. Levick graduates from London's St Bartholomew's Hospital then immediately applies to join the Royal Navy. His first significant posting is to the Mediterranean where he becomes interested in the means of transmission of the bacterial disease brucellosis, which causes fever-like symptoms and is sometimes called Mediterranean Fever because of its prevalence in the area, where it is typically transferred to humans from goats and sheep.

It is here that he gives a glimpse of his character, both as a risk taker and someone who is actually cut from the same cloth as Amundsen; someone not inclined to shy away from hardship. Levick demonstrates that brucellosis is not transmitted in the urine of infected patients by drinking the urine himself. He also allows himself to be bitten by a mosquito after it has gorged itself on the blood of a very sick patient with a temperature that hovers around 103°F. Although he had earlier contracted malaria from mosquitos when in Sardinia, neither he nor his coconspirators contract brucellosis via the mosquitos. Levick and a colleague publish the results of their interesting, if ill-conceived, experiments in the *British Medical Journal* of 1905.

Yet in his observationally driven, experimental approach to disease transmission, there are the hallmarks necessary to conduct the first systematic study of penguins.

<hr />

It is October 23, 1977, and I am about to begin my own systematic study of penguins. The Iroquois helicopter swivels around on its approach to Cape Bird and I imagine I can see the beach at Cape Tennyson where Borchgrevink and Jensen were nearly drowned by a calving glacier. We swing over the large glacier that runs into the sea from the Mount Bird Ice Cap like a giant white tongue, ending in sheer 150-foot-high cliffs of ice. Suddenly, dramatically, a bare half circle of land lies before us, cradled by the ice cap. In a landscape of endless white, this expanse of brown comes as a shock; its barren, guano-stained soil a testimony to the importance of this ice-free area for the sixty thousand Adelie penguins that come here each summer to breed. As far as I can tell from the helicopter, the Cape Bird penguin colony is deserted of penguins.

The absence of penguins means that we can land the helicopter beside a green hut, which sits atop a ridge at the western end of the colony. Any later in the season and, in order to avoid disturbing the penguins, we would have had to land down on the beach and carry our supplies and fuel several hundred yards up the steep path to the hut.

The small hut that will form my home for three-and-half months is uninspiring. At its back there is a door to a tiny room without windows that functions as both a refrigerator for storing our food and an airlock to retain heat in the hut proper. The main room of the hut is too small to swing a penguin, let alone a cat. Six bunk beds make up two sides of the cabin. There is a tiny table with seats for two. A little alcove forms a tiny kitchen, with a bench just large enough to take a Primus stove and room for little else. There is a tiny window, but it is really just a piece of clear plexiglass screwed to the walls of the hut. Unlike Borchgrevink's hut at Cape Adare, which was built in 1899 and even then contained a double skin of wood with papier-mâché for insulation, this hut was built by the University of Canterbury in the 1960s as a temporary shelter and was made simply of three-quarter-inch plywood without a skerrick of insulation. Ice can be seen climbing up the inside walls of the hut and, even with the heater on, it will retreat but not fully disappear. The heater itself is housed in a small adjacent room that makes the main room seem huge by comparison. This space, about six feet wide and a dozen long, will act as our laboratory, laundry, and bathroom for washing ourselves. Using the toilet necessitates a trip outside to a small lean-to at the back of the hut, barely big enough to house a forty-four-gallon drum that has been sawn in half and has a toilet seat plonked on top. Strictly for number twos. Down a path, in a gulley fifteen yards from the hut, there is an orange plastic cone sitting in a pipe that goes into the snow: it forms a crude urinal that transports our urine under the snow and, presumably, down to the beach, out of sight and out of mind.

The helicopter takes off with a blast of wind and snow from its blades, leaving me and Max, the field assistant assigned to me by the New Zealand Antarctic Research Program, alone at the hut. Three companions from the University of Canterbury will join me in a few weeks, but until then, safety protocols dictate that there needs to be a minimum of two in any field party. Max is an experienced mountaineer; a man of actions but few words. He is my equivalent of the Laplander that accompanied Borchgrevink: a human safety device. He is there to ensure that I survive, not to help with the science.

I leave Max to figure out how to crank the heater up and head off to inspect the bare penguin colony. The penguins, which migrate north during the Antarctic winter, have yet to arrive. That is, with the exception, as it turns out, of a dozen individuals, almost certainly all males, which to my delight I find waddling up the beach and over the open areas of snow and guano-stained earth looking for their nest sites. Cape Bird is about 435 miles south of Cape Adare and, consequently, the Adelie penguins arrive here to breed later than those such as the one shown to the dying Nicolai Hanson.

Like Hanson, I am delighted by the sight of the first penguin. I am decked out in my padded yellow survival clothes with thick-soled mukluk boots. The first penguin I encounter comes up to my knees, a perfect picture of contrasts. Black-and-white. To the penguin, I must look like a giant yellow blob, a man mountain, if not a man of the mountains like Max.

It is love at first sight. At least for me. Until that moment, penguins represented nothing more to me than a ticket to the Antarctic. Yet, there is something in the attitude of this first penguin I encounter, more so than anything cute about its looks, that leaves me smitten. The penguin is so determined, so gritty, and as at home in this cold place as I am a stranger. I stand in its path but it is not afraid of me; nor will it alter its course because of my appearance. It stops, stretches out its flippers, fluffs out the feathers on its chest, and shakes its head. Drops of a clear kind of penguin snot fly out from the tip of its beak. But it does not retreat nor bow its head. If I represent a threat, it does not care; it seems to care only about one thing: the route along which it has chosen to move; the route that will presumably lead to breeding and success in an environment where I would flounder in minutes were it not for all the technology I wore in the form of my yellow clothing and big boots. Everything about the penguin says that it belongs here and I do not.

I move aside. It is me who bows. I bow to this little creature newly returned from the sea in its starkly colored outfit; the white feathers on its chest gleaming in the late evening sunlight. At that moment a fire is lit inside me that has yet to go out; a desire to understand everything about this black-and-white penguin and the world in which it lives. That moment is every bit as significant for me as the one that had lit a fire inside the seventeen-year-old Roald Amundsen the day he watched Fridtjof Nansen travel up the fjord to Kristiania. I could have described it similarly:

All my boyhood's dreams reawoke to tempestuous life. For the first time something in my secret thoughts whispered clearly and tremulously: if you could make an Antarctic penguin biologist!

It is while at the Fram Museum in Oslo that I come within touching distance of Roald Amundsen's boyhood dreams or, rather, the vehicle for potentially realizing them. An underground passage takes me to another building containing another wooden ship—although this one is shockingly small, hardly deserving of the designation "ship" at all. I walk down its length: it takes me just twenty-seven paces, but it is its narrowness that makes it seem so small, so sleek; just five or six paces across its width. It is quite strikingly colored in dark green, red, and black, with a rudimentary looking mainsail and a long bowsprit that can take three head sails. At its rear there is a propeller that looks ridiculously small, and yet what it represents is huge: onboard I find the room housing the engine that powers it. It is the first motor to have ever been put in a Norwegian ship.

The *Gjoa* has a definite elegance to it: if this were a person, I would liken it to a ballet dancer, lean and agile. I can see with my own eyes why Amundsen chose this little ship to attempt to "make the North West Passage" in responding to his own secret thoughts.

It is September 12, 1903. Roald Amundsen moors the *Gjoa* in a natural harbor on the southeast coast of King William Island in the Canadian Arctic, about halfway between the Atlantic and Pacific Oceans, about halfway to completing the oft-attempted but never-completed North-west Passage. It is fifty-seven years to the day since Sir John Franklin and the *Erebus* and *Terror* became beset in ice not so very far away as they attempted the same. Amundsen christens the harbor Gjoahaven. He cannot know that on the opposite side of the island, in Terror Bay, sits the *Terror* itself—it will take another 113 years before it is found—while its sister ship, the *Erebus*, also lies nearby.

The Northwest Passage is not Amundsen's only goal. One focus of his expedition is to get to the Magnetic North Pole. Amundsen calculates that it is only some ninety miles from Gjoahaven, hence, this seems like the perfect place to winter over and stage a sledging trip to the Magnetic North Pole the following spring. The harbor is protected, the surrounding land is not too steep, there are two streams bringing fresh water, and there are ample reindeer and geese for food—at least before the winter sets in.

As it turns out, Amundsen will spend two winters at Gjoahaven. Yet there is an upside in all this for a man intent on being a polar explorer. During his two-year encampment on King William Island, Amundsen befriends the local Inuit, a tribe known as Netsiliks. From them, he learns firsthand how to build igloos, how to use dogs, and the importance of eating meat to ward off scurvy. It all reinforces the lessons he experienced on the *Belgica*: that being well prepared is the only means to ensure survival and success in the world's harshest and coldest climates. It is the Netsiliks who show him how to do that.

What the Netsiliks do not show the man who professed on the *Belgica* to prefer ice to women is anything about the ways of sex. However, some of his men do fraternize with the Inuit women, despite their leader admonishing them not to do so, and even father children.

Even so, in the oral traditions of the Netsiliks, a line of them are sired by the tall Norwegian with the big nose.

Geir Kløver, the director of the Fram Museum, is as straightforward and no-nonsense as it is possible for a person to be. I get the impression that it pains him to smile. However, his oval face, topped with short, cropped hair, cracks into something approximating a grin when I ask him about Amundsen's purported Inuit children. He is adamant that there is no evidence for them and that it conflicts with Amundsen's admonition to his men to avoid fornicating with the natives. Underscoring all this, DNA evidence has revealed absolutely none of Amundsen's blood in the line of Inuit supposedly descended from him.

In September 1903, as Amundsen settles into a winter in Gjoahaven, alone in his bed, to await his attempt to get to the Magnetic North Pole, oddly enough, at the other end of the world, Robert Falcon Scott is at that moment preparing to set out on an attempt to reach the Magnetic South Pole. After Scott has been forced to spend a second winter in Antarctica because the *Discovery* has remained stuck in the ice at Hut Point, he and six others depart for the Magnetic South Pole, man-hauling their sleds. During the seventy-three-day trip they fail to reach their goal, but Scott and two of the others become the first humans to get up onto the Antarctic Plateau. Their critical achievement is that they make reasonably good distances each day by pulling their sleds: it is enough to cement Scott's aversion to using dogs and to reinforce his belief that the best way to travel in the Antarctic is by man-hauling.

The previous summer, while Scott, Wilson, and Shackleton had been trying, in vain, to get to the South Pole, others from his expedition had made a trip to Cape Crozier at the eastern end of Ross Island. There they discovered an Emperor penguin colony. Consequently, while Scott is heading off in one direction to seek the Magnetic South Pole, Wilson puts on his hat as the expedition's zoologist and heads in the opposite direction to Cape Crozier to study the penguins.

As well as being the site of a large colony of Adelie penguins, Cape Crozier is also the most southern breeding place in the world for Emperor penguins, the largest of all penguins. More striking to Wilson than the size difference between the two species, however, is his observation that while the Adelies are sitting on eggs and taking advantage of the relatively benign conditions of the Antarctic summer in which to breed, the large size of the Emperors' accompanying chicks—which are almost ready to fledge—suggest that the Emperor penguins must breed, unbelievably, during the horrifically cold and completely dark Antarctic winter. As improbable as that seems, Wilson is forced to accept the evidence before him:

. . . we find the Emperor penguin hatching out its chicks in the coldest month of the whole Antarctic year, when the mean temperature for the month is eighteen degrees below zero, Farenheit, and the minimum may fall to minus sixty-eight, I think we may rightly consider the bird to be eccentric.

~~~

Penguins were proving to be eccentric in more ways than one.

It is 1980, and I am at the University of Alberta in Edmonton, Canada, where I have gone to study for a PhD. While there, I am simultaneously writing up the results of my penguin research for a master of science degree from the University of Canterbury—the research that I had come up with at the last minute to get me to Antarctica.

I had begun my three-and-a-half-month stint in Antarctica by monitoring the penguin nests in six breeding groups—subcolonies—of the Adelie penguins at Cape Bird. Up till then, studies had determined breeding success and failure by monitoring the penguins' nests every five days or so. If the eggs or chicks in a nest disappeared within a five-day interval, this was typically put down to either predation by skuas or loss due to unknown causes. Skuas are large gull-like birds that set up territories around penguin colonies and try to pilfer as many eggs and chicks as they can, instead of going to sea in search of their usual diet of fish.

In contrast to those previous studies, I mapped out each nest site and inspected every nest and determined its contents every day. I would stand outside the subcolony and wait until I could see the eggs or chicks within each nest. I carried a long bamboo pole so, if necessary, I could gently raise the penguin's bum to peak at the eggs. Typically, this technique was not needed once the chicks hatched as they were more easily seen. I also set up a tent on the ridge above the subcolonies from which I could make additional observations. And I could also monitor the nests with binoculars from the window of the hut's lab itself.

Through these much more intensive observations, I was able to ascertain accurately when predation occurred. However, I also saw that in many cases, the eggs were simply abandoned by a parent bird before being taken by skuas.

It is while analyzing my data in Edmonton that I begin to appreciate something new, even eccentric, about the penguins. Yes, predation by skuas is a major cause of breeding failure but, in many ways, the penguins are their own worst enemies. Almost as many eggs and chicks are lost because penguin pairs fail to coordinate their nest relief.

A female Adelie penguin will typically lay two eggs, three days apart. However, the eggs are large and they need to be incubated for just over a month before the embryos have developed enough to hatch as chicks. Penguins must get their sustenance from the sea, yet they must breed upon land. Consequently, while parents are in attendance at their nests, they are on a diet; they must fast.

An Adelie penguin at the start of the breeding season weighs in, fighting fit, at around eleven pounds. As one parent must constantly be on the nest to keep the eggs and young chicks warm, not to mention to protect them from the ever-vigilant skuas that are after a cheap meal, there is no way that an eleven-pound bird can manage to do all the incubating by itself; no way that an eleven-pound bird can go without food in the Antarctic for the entire incubation period. And that does not include the additional two weeks they are on shore for courtship before even starting incubation. Consequently, the only way for Adelie penguins to breed successfully is for pairs of penguins to coordinate their nest attendance patterns so that one is always on the nest while allowing the other to be away at sea feeding.

Analyzing my data shows that the male Adelie penguins arrive at the subcolonies to breed usually a little before the females. Males and females engage in a courtship period that lasts about twelve days and involves a fair amount of the lovemaking that so excited Borchgrevink's men. Once the female has laid both eggs, however, she takes off to sea, leaving the male to incubate the eggs for about two weeks. When she gets back, it is the male's turn to go to sea to feed for an extended period, which may also last for about two weeks. Thereafter, the pair takes turns on the nest at much more frequent intervals, typically trading places every one to three days. When the chicks are around two to three weeks old, they clump together with other chicks in their subcolony, forming crèches, at which point both parents can be away from the nest at the same time to get food for their chicks and themselves.

Survival analysis is a statistical technique used in medical research to measure time to death and identify critical periods of risk for patients. It has never previously been applied to penguins, but the detailed nature of my data allows me to use it to examine the causes of breeding failure in my penguins. This analysis shows that the risk from skuas is fairly constant throughout the life of an egg or young chick, but there are two periods of extraordinarily high risk of death due to other factors. When eggs are around eighteen days of age, the risk that a parent will desert them is enormously high. Similarly, when chicks are four to six days of age, there is a really high risk that they will starve to death.

As bizarre as that sounds, it makes perfect sense when matched with my detailed records of the nest attendance patterns of pairs at each nest. Normally, a female takes about two weeks at sea after egg-laying, but if she does not return in time, her male partner will have little in reserve to continue fasting and incubating. A fasting male loses just under two ounces in body weight each day, mostly through burning its fat reserves. From another facet of my work, I have shown that about thirty days without food is all that most males can tolerate before their fat reserves are reduced to such low levels that they have go to sea to feed themselves or risk their own survival. Given that the average courtship period for the males was twelve days, this leaves about eighteen days that they could sustain incubation without being relieved. Indeed, my nest attendance records reveal that the high risk posed to eggs at eighteen days of age is a consequence of their mothers not returning from the sea in time to relieve their fasting and, by then, very skinny fathers.

However, penguin offspring still have every reason to fear their fathers as much as their mothers. Given that a female has just replenished her fat reserves when taking her turn incubating, she is unlikely to be troubled by any extra fasting needed to cope with a tardy male, but a big problem arises because the chicks will hatch when the eggs are about thirty-three days old. If the mother has not been relieved, she will have no fresh food in her belly to regurgitate to her hungry chicks. I discover a previous study, carried out in the 1960s in a time before ethics committees scrutinized research proposals: it shows that penguin chicks deprived of food can survive unfed for four to six days by utilizing the remains of the yolk sac they have retained from being an embryo in the egg.

In other words, the huge risk to the survival of Adelie penguin chicks at four to six days of age comes about because they have never been fed since the time they hatched. Indeed, my nest attendance patterns show exactly that: chicks starving to death at four to six days of age do so in nests where their fathers have not returned from feeding at sea to relieve their mothers.

Perhaps the craziest, most eccentric thing I discover that winter, as I sit at my desk in the redbrick Biological Science Building, poring over numbers written in pencil on graph paper with a calculator at my fingertips (this is in a time before personal computers), is that successful pairs of Adelie penguin parents have complementary nest attendance patterns. That is, if a female is away for a longer than normal time while the male takes the first incubation duties, he will go away to feed for a shorter than normal time, even though, by all rights, because he has fasted for longer and used up more of his fat reserves, he might be expected to go for longer.

This raises two questions for me. Could it mean that successful penguins are responding to some sort of internal timer that tells them when their eggs are nearly done and that it is time to head back to the nest with food for their chicks? And, irrespective of that, shouldn't it give an advantage to any aspect of their mating behavior that ensures penguins are more likely to end up with a complementary partner?

I look out my window at the snow-covered ground and the dark, leafless trees lining the Saskatchewan River. It is not quite as black-and-white a world as that of the penguins, but the wan winter light of Edmonton seems to be reflecting the world of my thoughts as shades of gray. I have uncovered more questions than I have found answers.

Perhaps all that fighting among the mating penguins that Borchgrevink and his men enjoyed so much has a purpose after all?

My doctoral thesis at the University of Alberta is on ground squirrels not penguins. One of my advisors is in the anthropology department and that brings to my attention one of her colleagues: forensic anthropologist, Associate Professor Owen Beattie.

Beattie is somewhat pudgy, with longish, lank hair and thick-lensed, oversized glasses. In that sense, he is like hundreds of other academics on the University of Alberta campus that are obsessed with their academic work to an extent that they are not with their personal appearance or exercise. Where he stands out, and the reason he shows up as such a large blip on my radar, is the subject of his research.

Beattie announces an audacious plan to go to King William Island in the Canadian Arctic to look for the bodies of the men who were part of the Franklin Expedition. It seems too weirdly coincidental that I should be there writing a thesis about penguins in Antarctica at a place named after the first lieutenant of the *Erebus*, while this man down the corridor from one of my supervisors should be searching in the Arctic for people who had sailed on the same ship. Beattie's plan is to apply forensic techniques to any body parts and artifacts he can find in order to determine how and why the men of the expedition died.

In July 1981, as I am completing my last season of fieldwork on the ground squirrels, Beattie and his team traverse the coastline of King William Island looking for signs of the crews from the *Erebus* and the *Terror*. Near Booth Point, less than twenty miles from Gjoahaven, they discover skeletal remains, along with artifacts that they are able to use to ascertain, unequivocally, that the human bone fragments are those of a member of the Franklin Expedition. Forensic analysis reveals cut marks on the bones, which corroborates the oral histories of the local Inuit that say the men who abandoned the icebound *Erebus* and *Terror* were reduced to cannibalism in what was, in the end, a futile bid to stave off starvation. Most intriguing of all for Beattie, however, is that his analysis of the bones reveals extraordinarily high levels of lead.

Could it be that the men were killed more by lead poisoning than the cold or a lack of food?

The problem for Beattie's hypothesis is that lead levels in the bones could be a consequence of a lifetime's exposure to lead rather than something that occurred during the Franklin Expedition itself. What he really needs is soft tissue from the dead crew members to analyze. As it happens, marked graves of three of Franklin's crew have been found north of King William Island on Beechey Island.

It is 1984. The PhD is behind me, and just as I am contemplating a return expedition to Antarctica to continue my penguin research, Beattie leads an expedition to Beechey Island and exhumes the bodies of John Torrington, John Hartnell, and William Braine. Their soft tissues have been remarkably well preserved in the permafrost of the Canadian Arctic. Beattie and his team are able to identify scurvy and lead poisoning as causes of death. Furthermore, the lead in the bodies matches exactly the lead used to solder and line the empty tins of food from the Franklin Expedition, which Beattie has been able to recover nearby.

A lack of vitamin C, plus food contaminated by lead were, therefore, the reasons why the men of the Franklin Expedition failed to survive in an area where the Inuit not only survived, they thrived.

It is August 13, 1905, and Amundsen, who has been thriving on King William Island for almost two years learning the ways of the Inuit, sails out of Gjoahaven.

The previous year he had set out with sleds and dogs to get to the North Magnetic Pole. Eventually, Amundsen reached the position recorded seventy-four years earlier by James Clark Ross when he became the first person to reach one of the Earth's magnetic poles, only to discover that the pole had moved to the north. While Amundsen had thereby become the first person to prove that the magnetic poles shifted, it frustrated him greatly that he was not able to get to his goal, the current position of the Magnetic North Pole.

Yet, he learned many valuable lessons in the process of trying, especially from a group of Netsilik Inuit he encountered along the way, about how to apply ice to the runners of sleds to make them glide irrespective of the temperatures or the snow and ice conditions; about how to wear caribou skin clothing loosely so as not to sweat (the enemy for those undertaking polar exploration and wanting to stay warm); about how to manage dogs for hauling sleds efficiently; and about how superior dogs were to what he called the "futile toil" of man-hauling. He learned too that dogs can be sacrificed and fed to their own: they will cannibalize their companions with relish. And, further, that dog meat can make a suitable meal for men:

> *We ourselves tried some substantial steaks and found the meat excellent.*

Robert Falcon Scott and Roald Amundsen have proven themselves to be polar opposites in every sense of the words. While encamped at opposite ends of the Earth in order to attempt to get to their respective magnetic poles, they have drawn exactly opposite conclusions about the best way to travel in polar regions. One of them must be wrong.

As Amundsen takes the *Gjoa* out into Simpson Strait, they pass Hall Point where there is a grave site of two men from the Franklin Expedition. The failed expedition that had so inspired the young Roald is now tangibly close. To honor the dead members of the Franklin Expedition, Amundsen hoists the *Gjoa*'s flag and they "went by the grave in solemn silence."

Thirteen days later, on August 26, 1905, they sight another ship coming toward them. Amundsen has done the unthinkable to all but himself. He has sailed the Northwest Passage.

> *The North-West Passage had been accomplished—my dream from childhood. This very moment it was fulfilled. I had a peculiar sensation in my throat; I was somewhat overworked and tired, and I suppose it was weakness on my part, but I could feel tears coming to my eyes.*

CHAPTER SIX
LOST OPPORTUNITIES

I t is October 1984, and, arguably, I have accomplished my boyhood dream too: I am sitting in a Starlifter jet once more, heading to Antarctica for the second time. Admittedly, it has not been as difficult for me to obtain my dream as it was for Amundsen, yet I can now justifiably claim to have also fulfilled its modified aspiration, the one I added seven years earlier on an evening at Cape Bird when I met my first Adelie penguin. I am going back to Antarctica as a fully fledged Antarctic penguin biologist, not someone who views penguins as a substitute for seals or a ticket to the ice. I am now Dr. Davis and I have a masters thesis and a couple of publications about penguins under my belt. I am leading a team of five researchers on a three-month study to examine the breeding behavior of Adelie penguins.

After we land in McMurdo and complete our survival training at New Zealand's Scott Base, we take a helicopter out to Cape Bird to begin the study with no less of an audacious goal than that of observing a subcolony of Adelie penguins around-the-clock for an entire breeding season. Nothing like it has ever been attempted before with any bird.

We are only able to contemplate this because of the unique set of circumstances afforded to us by the Adelie penguins.

They breed from late October to late January, in the Antarctic summer when there is twenty-four-hour daylight. They are relatively large birds that nest close together on open ground, making them easily observable. Further, they are unafraid of us, having been virtually unexposed to humans, allowing us to set up an observation post near to their nests without affecting them going about their normal business.

The only difficulty is that they all look alike, making identification and observation of individuals impossible. I solve this problem by using a technique I had employed when studying the behavior of ground squirrels in Canada for my PhD, a species where all the individuals also look alike. In that instance, I caught the squirrels and painted a large identifying code that consisted of a letter and a number on their backs using Lady Clairol blue-black hair dye. Such hair dye is never going to work on the black backs of the penguins, but the principle of marking them should. At the start of the season, before the penguins even begin to start breeding, we catch all eighty-three penguins in the subcolony. We weigh them, measure them, put an individually numbered metal band around one flipper and, importantly, paint a coded combination of a letter and number on their backs using white enamel paint. Over the course of the penguins' breeding season, the paint will wash off and fade, eventually disappearing altogether when they molt their feathers at the end of the breeding season. In the meantime, it gives us the perfect way to quickly and accurately identify all individuals in the subcolony.

We set up a tent as an observation hide. Not to obscure us from the penguins—because, truly, they couldn't give a damn about our nearby presence—but to shield us from the worst of the Antarctic weather when observing the penguins through all twenty-four hours of each day. Two blizzards with one-hundred-mile-per-hour winds within the first few days soon put paid to that. The first destroys the tent; the second blows away our other tent. Thereafter, I opt to use the wooden crate that our generator had been transported in. It is too small for an observer to get fully inside, but it breaks the wind from tearing at the upper parts of our bodies and, thereby, provides a modicum of protection. Nevertheless, it can be bloody cold sitting out there for hours at a time, so we always dress in full survival gear and take with us a thermos of hot coffee or hot chocolate and some comfort food.

Our observation post is above the subcolony, and from there we have a perfect view of all the nests. We record the behavior of the eighty-three individual penguins present in the subcolony throughout the breeding season.

As masochistic as it might seem to sit outside in the Antarctic for three continuous months, the results are worth it. The first thing we observe is that these birds, which have been presumed to be largely monogamous, are anything but. They turn out to be what I call "serially monogamous," in that they seem to have only one partner at a time, but switch partners whenever a better opportunity affords itself. During the courtship period, about one-third of all the birds copulate with two partners, and some even do so with three.

Throughout that summer, as I sit in the generator box overlooking the colony, flanked on one side by the glacier that is the Mount Bird Ice Cap and on the other by the Royal Society Mountain Range, sitting across the ice-encrusted waters of McMurdo Sound, I contemplate why that should be. It is not like these are one-night stands or a bit of sex on the side. These start out as the penguin equivalent of marriage. Why commit to one bird then a few days later turn your back and commit to another? Biologically, let alone morally, where was the literal fucking sense in that?

I stare out at the rounded shape of Beaufort Island sitting some fifteen miles off Cape Bird, which in the super-clean Antarctic air looks so much closer. I stare out at the Ross Sea with its shifting cover of pack ice and icebergs that is constantly changing from one glance to the next. Sometimes its waters heave with minke whales, sei whales, and most exhilarating of all for me, killer whales. And though I never quite land a satisfactory answer to these questions about the apparent immorality of the penguins, I get a sense that the answer must somehow lie there in the blue waters of the Ross Sea before me; that somehow the requirements for a penguin to breed and feed so far south creates the circumstances where all the perceived wisdom about monomorphic seabirds being monogamous can be thrown out the window or, in my case, tossed out of the generator box.

It is January 22, 1985, near the end of the breeding season and instead of taking our helicopter directly back to Scott Base, we go via Cape Royds in order to census and measure the breeding success of a small colony of Adelie penguins that nest even a little closer to the South Pole than do those at Cape Bird. In fact, they nest closer to the Pole than any other birds. Even so, we discover they have managed to rear 3,457 chicks to fledging age this season.

Cape Royds also has another claim to fame: it is where Ernest Shackleton built a hut from which he made his own attempt to get to the South Pole.

⸺

In 1907, Ernest Shackleton is determined to prove himself "a better man than Scott." He launches his own expedition to get to the South Pole. Rober Falcon Scott is livid. Scott demands that Shackleton promise to steer clear of McMurdo Sound and the Ross Island area of Antarctica, which he regards as his "own field," as he puts it.

Initially, Shackleton tries to be true to the public undertaking he has given to Scott, albeit reluctantly, to stay away from McMurdo Sound. He heads, instead, for the inlet Borchgrevink had discovered in the Ross Ice Shelf and the nearby Balloon Bight, which, in a sign of the pique that has now infected his relationship with Scott, he refuses to call by Scott's name, instead calling it Barrier Inlet. However, when he gets there on January 24, 1908, he discovers that the Ross Ice Shelf, which is really an enormous moving glacier, has calved off a huge section of ice that has taken away both Borchgrevink's bight and Scott's nearby one, thereby destroying the easy access to the ice shelf and the virtual highway to the South Pole that he had glimpsed from the balloon six years before. In their place, a large bay, or indentation, has formed in the edge of the ice shelf where whales abound. Shackleton renames the area the Bay of Whales, but he is not about to set up camp in such an unstable area nor one without such easy access as it had enjoyed before. He makes a tentative push eastward in a bid to get to King Edward VII Land, but the path of his ship, the *Nimrod*, is immediately blocked by a dense concentration of pack ice. Shackleton, just as quickly, decides to turn the *Nimrod* around and head westward.

Whether intentionally or by dint of conditions that Shackleton said left him no choice, Shackleton ignores his public undertaking to Scott and sets up his base at Cape Royds on Ross Island.

———

I am standing in front of the door to Shackleton's hut three decades after my first visit. On that occasion, the experience of going inside Shackleton's dimly lit hut had proven as close to a religious experience as I had ever had. It had seemed to me more of a shrine to an extraordinary man than it was a receptacle for the very ordinary things it contained: cans of meat, old boots, a broken sled, a broken promise, and the echoes of hearty comradeship. This time, however, I have come not seeking Shackleton but the ghost of Murray Levick.

To the detective in me, it is becoming clear that I need to reconstruct the sequence of events that took Levick to Antarctica in order to understand how he should become the world's first penguin biologist. I have a hunch that there are important clues to be found by examining Shackleton's Nimrod Expedition of 1907–09 and this hut in particular.

Snow is falling lightly and the clouds are dark and bruised. The piles of wooden expedition boxes and bales of hay outside the hut have been tidied up somewhat and there is a new skin of some light-gray, rubbery-looking material that has been applied to the roof by conservators, but otherwise the hut looks the same as it did three decades earlier, or, indeed, when it was built more than a century ago.

Ostensibly, it was Robert Falcon Scott—even more so than Borchgrevink, Amundsen, Nansen, or anyone else—who was responsible for

Murray Levick coming to Antarctica and ending up at Cape Adare with nothing to do. However, I believe that the influences that led Levick to study Adelie penguins and, especially, their sexuality, are too many to place solely at Scott's door. As I stand there brushing the snow and volcanic grit from my boots in preparation for going inside, it is apparent that some of the responsibility for turning Murray Levick into the world's first penguin biologist can be laid at this door: the door of Shackleton's hut at Cape Royds.

I think back to Christmas 2007. In the days leading up to Christmas, a report emerged that two chalk sketches of penguins on blackboards had been found in a basement at the University of Cambridge. Given the ephemeral nature of chalk on blackboard, it seemed amazing that they should have survived at all because they were dated from 1904 and 1909, making them 103 and 98 years old. Even more amazing were the identities of the artists. The first was signed by Robert Falcon Scott and came from a public lecture he gave at Whitworth Hall in Manchester on December 1, 1904, less than three months after getting back to England from the Discovery Expedition. The second was signed by Ernest Shackleton, from a public lecture about his Nimrod Expedition, given at the same hall five years later.

Aside from the novelty value of the sketches, the media commentary focused on cheap asides along the lines of both men being better explorers than they were artists. For me, that was not the striking thing at all about the drawings. In fact, to be fair to both men, they are actually pretty good representations of Antarctic penguins to have been drawn quickly on a blackboard while giving a lecture to an audience that cannot have ever seen such animals.

No, for me, the striking thing is that Scott chose to draw an Emperor penguin, while Shackleton drew an Adelie penguin. That single act spoke volumes about where each's affinities lay concerning penguins.

Scott's zoologist and friend, Edward Wilson, had expressed disgust about the prospect of living at Cape Adare among the smell of all the Adelie penguins. While Scott had been out on the Antarctic Plateau man-hauling sleds toward the South Magnetic Pole, Wilson had been at Cape Crozier admiring the audacity of the Emperor penguins to breed there during the Antarctic winter. Scott chose to moor the *Discovery* at Hut Point and used that as his base. Devoid of nearby penguin breeding colonies, most penguins whose wanderings took them to the ice edge around Hut Point were likely to be Emperor penguins. The Emperor penguin is pompous and grand, a perfect reflection of its name. It is large, beautifully colored, and it carries itself with a slow, stately kind of grace. It is easy to imagine that Scott would have been most smitten with Emperor penguins. They mirrored his own values; those of a man who prized the conquest of Antarctica when it was "more nobly and splendidly won."

By contrast, Shackleton chose to base his expedition beside a colony of the stubby, raucous, black-and-white Adelie penguins. While it was not the primary consideration at all in where he decided to build his hut—ice conditions prevented Shackleton from getting as far south as Hut Point in 1908—he nevertheless chose, to me at least, the most scenically attractive place in all of the Ross Sea area. His hut is backed by the active volcano, Mount Erebus, which rises over twelve thousand feet above sea level with an ever-present plume of smoke emanating from its crown. Across the ice-covered waters of McMurdo Sound, the view is of the magnificent Royal Society Range. The hut itself is nestled beside a small lake and about three thousand pairs of Adelies make the surrounding buttresses of volcanic rocks their homes during the summer. The Adelie penguins were not just Shackleton's neighbors, in a sense, they were his companions too.

It is 1908. Shackleton's Nimrod Expedition includes the biologist James Murray. Given Hanson's untimely death just as the Adelie penguins were beginning to arrive at Cape Adare during Borchgrevink's expedition nine years earlier, and given Scott's decision six years earlier to set up his base at Hut Point, miles from any penguin colony, Murray has the opportunity to become the world's first penguin biologist gifted to him on a plate. He is living in the hut at Cape Royds, literally yards from the nesting Adelie penguins.

Murray, however, is monumentally not up to the task. An insight into his perspective may be gathered from his reaction to Cape Royds itself.

> To the biologist, no more uninviting desert is imaginable than Cape Royds seemed when we made our first landing, and for long afterwards.

He describes, in detail, the rotifers and other invertebrates he finds in the ponds around Cape Royds, but his observations about penguins are as far from scientific as it is possible to be.

> There is endless interest in watching them, the dignified Emperor, dignified notwithstanding his clumsy waddle, going along with his wife (or wives) by his side, the very picture of a successful, self-satisfied, happy, unsuspicious countryman, gravely bowing like a Chinaman before a yelping dog—the little undignified matter-of-fact Adelie, minding his own business in a way worthy of emulation.

As a consequence, when the *Nimrod*, with Shackleton and his crew on board, finally leaves Antarctica at the end of the penguins' breeding season, in February 1909, and heads back to England where Shackleton shall give his lecture at Whitworth Hall, the opportunity to become the world's first penguin biologist still sits begging.

Yet, there is one crew member on the *Nimrod* destined to have a major impact on Murray Levick: it is the young geologist, Raymond Priestley. Born in 1886, a full decade after Murray Levick, Priestley is just twenty-one when he heads to the Antarctic on the *Nimrod*, having taken a break from his studies of geology at the University of Bristol. Handsome, tall and lean, with tousled hair, Priestley looks more like the child of athletes and actors than the son of the headmaster of Tewkesbury Grammar School.

That athleticism, if not the good looks, comes in handy when he is in a party of five men that sets out to climb the northern slopes of Mount Erebus. When the men leave the hut at Cape Royds, the weather is fine and so they opt to travel light, carrying but a single three-man tent with them.

As with so much that goes wrong during polar exploration, it is not so much the conditions themselves but the lack of preparations to deal with them that creates the need for heroic acts that separate survival from death, even if barely so. In this case, an almighty blizzard strikes while the men are high up on a glacier. In their tent, even by squeezing themselves in, there really is only room enough for four. Priestley volunteers to stay outside, hunkering down in his sleeping bag. For three days the wind and snow rage, battering Priestley while he lies helpless and frozen in his sleeping bag, blowing him slowly down the glacier toward precipitous hundred-foot-high ice cliffs. Without anything to drink, he uses his fingernails to scrape tiny fragments of ice from the glacier surface, which he sucks. After nearly two days, one of the men from the tent crawls out to bring him some chocolate. By the time the wind abates enough for the others to leave the tent again to look for him, his feet are frostbitten. They bring him to the tent where he lies upon the others.

As soon as they are able, the men abandon the climb and head back to the warm and comparatively roomy hut at Cape Royds. These men have failed to accomplish what they set out to do and almost squandered at least one of their lives—all through being unprepared for the challenges that being in the Antarctic can throw at them. On the other hand, it marks out Priestley as being made of special stuff. If in the future Murray Levick should ever need a companion when circumstances turn dire, he could hardly hope for a better person than Raymond Priestley.

There is another member of the Nimrod Expedition, another geologist, who will prove himself as mentally and physically the equal of either Shackleton or Priestley. On March 10, 1908, a strapping young Australian, Douglas Mawson is one of five men who are the first to reach the summit of Mount Erebus. If Mawson thinks the climb is hard or dangerous in any way, it will prove to be neither—at least in comparison to what he still has waiting for him in Antarctica.

During the summer of 1908–09, Shackleton's party splits up into four groups: the Northern Party, consisting of three men including Mawson, head off with the aim of being the first to get to the South Magnetic Pole; the Western Party also contains three men, including Priestley and an Australian, Bertram Armytage; the Southern Party of four is headed by Shackleton, and it sets out with no lesser aim than to be the first to the South Pole. Finally, a party of five men remains at the Cape Royds Hut with biologist James Murray as their nominal leader in the absence of Shackleton.

Mawson and his two companions do what Scott was unable to do and they become the first humans to make it to the South Magnetic Pole, which is situated on the Polar Plateau in as isolated and unwelcoming a place as it is possible to be. Mawson and the others are taunted by death on so many occasions that it seems inevitable that if it is not this crevasse, it will be the next in which they fall into oblivion. There is also the almost complete physical and mental breakdown of the party's leader, which requires a reluctant but, as it proves, highly capable Mawson to take over. Somehow, he leads them all to the conclusion of their 1,260-mile journey over some of the worst conditions ever encountered by polar explorers at either end of the Earth. They manage to just make a fortuitous rendezvous with the *Nimrod*, having failed to reach the coast before the ship sailed by, searching for them. The first officer of the *Nimrod*, John King Davis, convinces the captain to go back to check an area where some grounded icebergs may have obscured their view at the base of the Drygalski Ice Tongue. There, behind the bergs, they find the newly arrived members of the Northern Party.

Priestley and his men also have a lucky escape. They enjoy a successful time exploring the region around the Ferrar Glacier but then, in a bid to ease the difficulties of the *Nimrod* to get to them because of the frozen sea ice around their rendezvous point at Butter Point, they camp on the sea ice. During the night, they awake to discover that the ice has broken out and they are floating on the Ross Sea: in all likelihood, the last bit of sailing any of them will ever do. For a day they are carried away from the shore by the wind, but at one point it turns and blows the ice floe back toward the shore. They calculate that the ice floe they are on might just catch an edge of the sea ice jutting out from the land and, in the briefest few seconds that it does so, they haul themselves and their sledge to the safety of the shore before the ice floe is driven north in what would have been their route to a certain death. Soon afterward the *Nimrod* arrives to pick them up.

Of all the parties, though, it is Shackleton's Southern Party that has it worst. Initially, they proceed south across the Ross Ice Shelf with four ponies—all that remain of the ten that had left Lyttelton on the *Nimrod* a year earlier. The ponies do not prove as resilient as the men and, one by one, they must be put down, until the very last of them falls to his death down a huge crevasse as they negotiate the Beardmore Glacier. The Beardmore Glacier is a 125-mile glacier that Shackleton has discovered, which acts like a very dangerous road from the barrier ice of the Ross Ice Shelf, through the mountains, to the Antarctic Plateau itself. The going is painful and painfully slow: the men haul their two sleds up *strastugi*, which are sharp-edged waves of ice formed by the wind. They fall, it seems, as often as they go up: down strastugi, down crevasses, until it seems like they can move no more. Their clothing is woefully inadequate as they walk into fierce winds that freeze their faces including the hairs inside their nostrils. They are getting weaker by the day: the exertion and dysentery from a terrible diet weakening each of them equally. Despite having meat from the horses and even the remaining feed intended for the horses, which they add to their soup—or "hoosh," as they call it—that they eat each morning and night, they are forced to cut rations to a few biscuits and two small bowls of hoosh each day. The four men are running out of food.

Finally, when within 112 miles of the Pole, Shackleton is forced to call a halt to their slog. While they still, somehow, have the will and probable strength to make the Pole itself, there is no way they would have enough rations to get back alive. Even at that point, their getting back is not a sure thing. When eventually they arrive at their food depot, the four men can barely walk and are close to death. One of them collapses and can go no further. Shackleton makes the decision to leave two of them there, while he and another make a dash over the last thirty-eight miles to Scott's old store hut at Hut Point where, according to Shackleton's written orders before the parties had gone their separate ways, there are supposed to be food supplies and a party waiting. There are neither. Yet after a cold night spent without sleeping bags (they had left theirs behind to travel more quickly), they light a fire to signal the *Nimrod*, which they hope is still somewhere in McMurdo Sound, and, sure enough, having already picked up the personnel from Cape Royds, the *Nimrod* is headed to Hut Point.

After a bath and some food, despite being physically knackered from his harrowing journey of 1,700 miles, walking to a point closer to the Earth's axis than has been attained at that time at either the North or the South Poles, Shackleton insists that it be he who returns with two others to get the remaining two men from his party.

Once all are safe and sound onboard the *Nimrod*, they head north for New Zealand, sailing past the hut at Cape Royds but not setting foot in there again.

It has stopped snowing for the moment. Yet the dark purple clouds swirling overhead suggest that there is more to come. There is a new sign screwed to the door of Shackleton's hut that proclaims, *This building and its contents are a historic shrine . . .*

It had been biologist James Murray, who having shared the summer of 1908–09 at Cape Royds with the penguins, had been last to go through that doorway. During that summer, while the other three parties were out risking life and a number of limbs in the name of exploration and science, he had nowhere else he needed to be, no other task to perform than to stay where he was, cheek-by-jowl with a colony of Adelie penguins. It seems flabbergasting to me that, as the expedition's biologist, he did not set about studying the Adelie penguins given the unfettered opportunity afforded to him.

Yet if nearly five decades as a biologist have taught me anything, it is that there are two types of biologists. One, to which I claim affinity, is attracted by the rawness and excitement of the great outdoors and especially those charismatic megafauna like big cats, wolves, bears, seals, whales, and yes, penguins. Danger and drama are constant companions for such scientists and, truth be told, a big part of the attraction too. The other type of biologist is generally happy indoors, looking down a microscope or prodding their biological subjects with some piece of equipment or another. Mostly, they are fascinated by small creatures where the biggest danger they face is spilling a hot cup of tea over their white lab coats. James Murray was one of the latter. The things that beguiled him most while at Cape Royds were the tiny rotifers he dredged from the bottoms of small lakes. He was fascinated by how hardy they were, conducting experiments to show that they could survive freezing or, conversely, really hot temperatures. All the while, outside the windows of the Cape Royds hut, a colony of knee-high penguins performed the perverse and debauched behaviors that characterize their mating, untroubled by Murray's eye any more than they had been by those of Hanson, Borchgrevink, or Wilson before him.

To the contrary, among the limited notes Murray wrote about the penguins, he opined:

> . . . *the Adelie appears to be entirely moral in his domestic arrangements.*

The inside of the hut is suffused with a soft dull light from a couple of small windows, desaturating further the room's already sepia-colored contents: a stove, some beds, items of clothing and boots, and stacks of various cans and goodies that brought such humor to the room over one hundred years earlier. The sign is right: it still feels more church or shrine than it does the refuge of men who could walk to within 112 miles of the South Pole or lie outside in a blizzard for three days and survive. I speak in whispers, if I speak at all. So completely have the room and its contents been preserved in the dry cold of Antarctica, and so artfully have they been protected by conservators, that to pass through that door is to travel back in time. Shackleton, Priestley, Mawson, and the others might have just stepped outside.

I stand alone in the hut. Something keeps me there, immersed in a childhood dream that is so tangible I can smell the blackened air that comes from cooking seal blubber on the stove; I can touch Priestley's sock as it hangs to dry, having done its best to protect his foot from frostbite as he hunkered down in the blizzard; I can feel the inside of Shackleton's sleeping bag with molecules of him left there still to shake my hand. If the spirit of Antarctica resides anywhere, it is here. I stand there waiting, absorbing everything the men have left behind. I know they will not be stepping inside again, but honestly, I would not be surprised to see them, lost as I am in an eerie world somewhere between trance and truth.

With a sigh, I take a final glance at the photo on the wall of King Edward VII and walk out the door, closing it behind me. Shackleton had done something similar before he had left on his attempt to march to the South Pole. In fact, the evening before he left, the sun had come through the window and spotlighted the photo of the king in profile—all whiskers and sternness—in a way that it had never done so before. Shackleton, a spiritual if not a religious man, took that to be a splendid omen for good luck when, on October 29, 1908, he went out the door of the Cape Royds hut and headed into the white nothingness that lay before him in order to seek glory for king and country, and, it has to be said, not a little for himself.

Four months later it would be biologist James Murray who, after a summer spent with the penguins would shut the door of the hut for the very last time, leaving the penguins still unstudied, before he boarded the *Nimrod* and departed to pick up the exhausted, near dead members of the Southern Party.

Neither James Murray nor Ernest Shackleton could have imagined that the next person to go through that door would be Murray Levick.

CHAPTER SEVEN

COURTSHIP

Murray Levick still had a number of steps to take before he could walk through the door of Shackleton's hut, and I have come to St Bartholomew's Hospital in London to try and connect the dots.

St Bartholomew's Hospital, or Barts as it is affectionately known, is Britain's oldest hospital. It was founded in 1123 by one of the courtiers of Henry I, William the Conqueror's son and the brother of Robert Curthose. If King Henry I is remembered for his fighting, it is another King Henry, known much more as a lover than a fighter, who has left his mark on Barts. I walk out of the stone archway that is the King Henry VIII Gate of St Bartholomew's Hospital, with a statue of the king himself standing atop the arch, a potent symbol for sexual libido. That seems especially appropriate considering that this is where Murray Levick, discoverer of the sexual excesses of penguins, had been tutored.

The buildings are all white ornate stone work and, save for a red telephone box and red double-decker buses, I imagine that little has changed from the time that Levick walked down that same street passing the same buildings with the beautifully rounded dome of

St. Paul's Cathedral at its end. Despite the grandeur of the buildings and the cathedral designed by Christopher Wren, the architect of the Rothschild mansion in Tring, I am left feeling short-changed by my visit.

I am struggling to find any evidence of what took Levick from a hospital to the Royal Navy to the door of Shackleton's Hut at Cape Royds in Antarctica. The one significant telltale sign seems to be his dedication to sport. He was fond of rugby, gymnastics, and rowing, all driven by his relentless pursuit of physical fitness.

It is in 1908, just as Shackleton is preparing to leave his hut to walk to the South Pole and six years after Levick walked out the King Henry VIII gate of Barts for the last time, that fate deals Murray Levick an especially auspicious hand: he is appointed surgeon aboard the battleship HMS *Essex*. It is commanded by none other than Robert Falcon Scott, who had been given command of the ship in January, eight months ahead of his marriage to Kathleen Bruce.

⸺

If I could, I would have set up my empty generator box from which to observe Levick and the other polar explorers, much as I did with the penguins. No doubt, I would have been similarly taken with their courtship patterns. As I am starting to appreciate, an understanding of what motivates the behavior of these males can be gained from observing their relationships with females. Just like the penguins.

It is September 2, 1908. Robert Falcon Scott is marrying Kathleen Bruce at the Royal Chapel in Hampton Court.

Scott is closer to being bland than he is beautiful. Yet his thick dark eyebrows emphasize his brooding eyes, taking attention away from his somewhat pointy ears and thinning hair. He holds himself upright and his stoic demeanor belies the troubled soul hidden behind those eyes. He is attractive enough that some, including his own family, regard him as something of a ladies' man.

When he was twenty-one, so the murmurs go, he had an affair with a married woman, Minnie Blanchard, during a stopover in San

Francisco. As was so often the way in such Victorian times, the Admiralty averted any scandal by promptly sending Scott back to England.

In the aftermath of the Discovery Expedition, Scott had become something of a celebrity. At a luncheon party in March 1906, he had met Kathleen Bruce, a sculptress of Scottish and Greek descent. Dark-skinned, bouncy, and flirtatious, Kathleen embraced a Bohemian, impulsive, fun-loving attitude to life. She is everything that Scott is not. She is a feminist before feminism will even have a name.

She also has a powerful and peculiar obsession: to bear a son who will become a hero. For that to occur, nothing will do quite so well as the boy having a father who is a hero too. Scott fits the bill perfectly. Already well known for his *Discovery* exploits, much greater acclaim surely beckons Scott in the form of the South Pole. Another expedition, to get to the South Pole this time, is not just a venture that Kathleen is prepared to support, she wants it more than anything.

After the wedding, Kathleen stays in London while Scott attends to his naval duties in Devonport and makes preparations for his next, as yet unannounced, expedition to the Antarctic. She does, however, summon Scott to London at the appropriate time in her menstrual cycle in order that he may impregnate her. It is a tactic that works brilliantly, and she writes to him early in 1909 to say that she is pregnant, barely hiding her ultimate motive:

> *Throw up your cap & shout & sing triumphantly, meseems we are in a fair way to achieve my end.*

Ernest Shackleton is as romantically different to Scott as it's possible to be. Big-shouldered and barrel-chested, he has a square jaw and thick dark hair that he parts down the middle. He is deep-voiced and charming; a rogue who spouts poetry.

While he was in the Merchant Navy, it seemed that every port held one girl, if not more, for him. Yet in June 1897, when he was just twenty-three, his erotic ramblings were brought, if not to a stop, then a pause: he met Emily Dorman, the friend of one of his sisters.

Tall and slim, Emily is blessed with big features: a prominent nose, big blue eyes, and large teeth that make for a particularly attractive smile. She is six years older than Shackleton and well educated in the arts. She shared his love of poetry and introduced Shackleton to the poems of Robert Browning. Thereafter, Browning became Shackleton's favorite too. He was smitten with Emily, but at first she paid him little heed, which only increased his determination to pursue her.

For Emily, the daughter of a well-to-do solicitor, Shackleton, a merchant seaman, was considered a little too beneath her class. It was the major reason why he signed up for Scott's Discovery Expedition to Antarctica: to impress Emily and prove himself worthy of her love, as well as being a way to accumulate some much-needed funds.

Shackleton asked Emily to marry him just before leaving for Antarctica with Scott in 1901. Perhaps fearing rejection, he sent a letter to her father requesting her hand in marriage only as the *Discovery* set sail. Her father replied in the affirmative, but by the time his answer caught up with the *Discovery* in New Zealand, he had died.

Shackleton confessed that he wanted to be engaged to Emily so as to thwart others who might be interested in her while he was away for years in Antarctica. As he put it in a letter to her, it is "a man's way to want a woman altogether to himself."

Despite being betrothed to Emily while in the Antarctic, such monogamous appreciations apparently did not go both ways. When invalided home by Scott, Shackleton had a romance with an adventurous young woman called Hope Paterson on the ship he took back to England from South Africa. He even wrote her a poem that concluded passionately:

> *Though the grip of the frost may be cruel and relentless its icy hold*
> *Yet it knit our hearts together in that darkness stern and cold*

Nevertheless, on April 9, 1904, after a courtship period lasting some seven years, Shackleton finally married Emily Dorman in Christ Church, near Westminster Abbey.

Yet, while on the Nimrod Expedition, he discovers, climbs, and names a mountain after Hope Paterson. Ostensibly, Mount Hope has been named because it lifted the spirits of the men upon reaching its summit, but subsequently he sends Hope a piece of rock from the mountain's top. It is mounted in silver metal with a small plaque that reads "Summit Mount Hope" and gives its height (2,785 feet) and geographical coordinates. Evidently, Antarctica is not the only land he has explored and laid claim to.

If nothing else, the rock, an echo of those prized by the penguins mating outside his hut at Cape Royds, is an indication that Shackleton is probably more penguin than he is a man of his word when it comes to marital fidelity. Among polar explorers, that is not a distinction he will have to himself.

The handsome hero Fridtjof Nansen is almost as famous for his affairs as he is for his conquests of Arctic regions. Perhaps the real surprise comes in July 1909. While Kathleen Scott is pregnant with her much wished-for son, the monk-like Amundsen, who had professed not to need women when on the *Belgica*, who has known sex—if at all—only in brothels and igloos, falls in love for the first time in his life at the age of thirty-seven.

His boyhood dream accomplished, having made the Northwest Passage, Roald Amundsen, like Robert Falcon Scott, has more ambitious polar plans. He is now setting his sights on the goal that had eluded his hero, Nansen: the North Pole. At a ball in Oslo, which Amundsen attends only out of duty that he might find backers among the elite of Norwegian society to support his new expedition, he meets a woman who had been in his class during his short-lived period at medical school in Kristiania University. Sigrid Castberg is married to a wealthy businessman. She agrees to help Amundsen raise money from her rich friends.

Where previously Amundsen had known only a kind of cold fusion between himself and the opposite sex, with this slim woman with the high cheekbones there is a chemistry of a different kind.

The Grand Hotel in Oslo is an ornate affair that sits at the corner of Karl Johans Street and Rosenkrantz Street. From its corner balcony

Fridtjof Nansen had received an ovation from the adoring crowd "amid the bunting and the cheers" the day he returned from Greenland, and so inspired the seventeen-year-old Roald Amundsen to become a polar explorer. Twenty years later, Amundsen has a different kind of exploration making his heart tremor with "throbbing pulses." He has been given a room at the hotel for his use when he is working in the city on his forthcoming expedition. One night, after dinner with Sigrid, ostensibly an innocent meeting about support for the new expedition, he goes to his room where, discreetly, she joins him sometime later. He is awkward and unsure, but she kisses him and the last piece of the ice shelf that has kept this man so stiff and removed from feminine involvements melts away. They make love on the bed and, for a moment, this normally aloof and calculating man becomes a bowing, immensely grateful male penguin in her hands. Amundsen is smitten.

I am out of my metaphorical generator box and it is now my turn to stay at the Grand Hotel. More than a hundred years after Amundsen, the hotel still manages to pull off its trademark style of understated opulence. There is marble where it needs to be, a dining room with a delightfully vaulted ceiling, plush chairs, and an abundance of tasteful artwork. The twin elevators, sheathed in brass, lead to what are now eight floors, many more rooms having been added since Amundsen's day. Yet the rooms on the first three floors overlooking Karl Johans Street are original. Sadly, there is no record of exactly which room Roald Amundsen and Sigrid Carsten used, but I imagine that it is the room I am in. Certainly the twelve-foot-high stud would not have changed, and the bay windows, while double-glazed now, are consistent with the original design. The chandelier might be a different one, but I am assured there would have been something similar in its place. The subdued gray walls and carpets are all brand-new and recently renovated, but I suspect that could Roald Amundsen have stood with me in that room, the only significant difference he would notice would be the big flat-screen TV. Not that it would matter to him. He had been intent on watching something else: Sigrid undressing. And if the bed then were a bit more uncomfortable than the massive king-size bed that is here now, I doubt that he would have noticed.

At almost exactly the same time that Amundsen is experiencing love for the first time, the same is happening to Douglas Mawson on the other side of the world in Australia. In August 1909, Mawson meets Francisca Delprat at a dinner party. The six-foot-tall, seventeen-year-old Dutch-born beauty has long, dark hair and an accent that Mawson finds as alluring as the rest of her. She had lived in Spain before emigrating to Australia with her family and goes by her Spanish nickname "Paquita," which means "free." If that were not enough for the gobsmacked Mawson, she is also keenly interested in both geology and him, and not necessarily in that order.

Murray Levick during this period, on the other hand, gives no sign of any form of engagement with the fairer sex. I discover some of his correspondence and it is exclusively to other males; exclusively work-related. If he were a penguin, I daresay that he would be one of those young unpaired males, setting up a nest on the periphery of a subcolony, collecting stones with which to attract a potential mate, only to have them stolen by a neighbor when he has his back turned.

⸺

From my time at Cape Bird, observing the real penguins around-the-clock throughout the courtship period in 1984, it seemed that while the penguins engaged in a bit of mate switching from one to another, there was nothing like the infidelities exhibited by a Peary, a Nansen, or a Shackleton. The penguins may engage in serial monogamy, having multiple partners sequentially in the one breeding season, but it seemed that they only ever had one partner at a time. And surely, once they had settled on a partner and produced eggs with it, conventional wisdom said that it should benefit them to stay together, if not mate for life. Divorce should be low. At least, that is what I anticipated.

It is late October 1985, and I have been bitten by a bug much more potent than the mosquito in Levick's brucellosis experiment: I am infected with the desire to go back to Antarctica again, to once more sit in my generator box observing the mating behavior of my penguins in the subcolony at Cape Bird.

Adelie penguins return to breed in the same subcolony from one year to the next. Even though the paint markings of my penguins had been lost once the birds had molted at the end of the previous breeding season, their numbered flipper bands can still be used to identify each bird.

The winter is a hard time for penguins. Once they have graduated from being chicks, it will remain the period of greatest risk throughout their lives. Many succumb during the annual winter migration, when the birds are forced to go north to avoid the worst that the Antarctic can throw at them. Usually, about one in every six penguins will fail to complete the journey each year, but during the worst winters as many as a quarter of them will die.

Yet even when both partners from the previous season have made it back safely to the subcolony, I discover a remarkable thing that season: about one-third of the erstwhile pairs do not reunite; they divorce. This seems extraordinary for a bird that we are repeatedly told mates for life.

My students and I adopt a familiar routine. We hunker down in the generator box, observing the sexual soap opera being played out in front of us.

Usually, although not always, the male arrives at the subcolony at the start of the new breeding season before his partner from the previous season. The nests of Adelie penguins consist of little more than scrapes in the hard ground that are lined with stones, although from one breeding season to the next, the stones get scattered around and so there is little to mark a particular nest other than the slight depression made by the penguins' scraping feet over many years. Remarkably, the penguins, once they have established a breeding site, go back to the same subcolony the next season and, in the majority of cases, the exact same nest site.

The male then sets about collecting stones to line the nest, either by getting them from outside the subcolony or stealing them from his neighbors' nests. An industrious male will build a nest that resembles a large dog's bowl, with the flooring and walls constructed of little stones. However, these "bowls" are not to keep water in but, actually, to let it out. Snow storms can occur at any time during the Antarctic summer, sometimes completely burying the penguins. Eggs are porous and the embryos they contain would soon die if the eggs were to sit for long in water. The stones allow the meltwater from the snow to drain away, leaving the eggs, which are tucked under the breeding bird's belly, nice and dry.

During the period of courtship, males—and especially unattached males—perform what has been called, rather incongruously, an ecstatic display. It starts with the male fluffing the feathers on the back of his head to form an angular crest rather than the smoother, rounder shape of an Adelie penguin when at rest. He then puts his head down, tucks his bill under one flipper, and emits a growling sound. Straightening up, he points his bill skyward, puffs out his chest and emits a series of stuttering grunts, all the while waving his flippers rhythmically as if a reminder of a time when his ancestors could fly. At last the grunts, which have been increasing in intensity, reach a crescendo and the call ends with a long, sustained, cry.

The male will then repeat the whole thing. And it is contagious. Other unattached males will call similarly. These are advertisements. Songs of Self. If these penguins were Ernest Shackleton, they would be reciting the poetry of Browning, for the purpose of their ecstatic calls is the same: to seduce would-be partners.

The males all look alike, to my eyes at least, and evidently to other penguins too. The calls of penguins, however, are individually distinct and, therefore, individually identifiable. Each penguin's call is like a fingerprint. My students and I investigate whether there are characteristics in the ecstatic calls of males that might make one male more attractive to females than another.

Adelie penguins usually breed for the first time when three to six years old. Virgin females prospect for their first partner by checking out unattached males when they are performing their ecstatic displays, oftentimes walking directly up to the calling males. What we find is that prospecting females prefer males with deeper voices: the Ernest Shackletons of the penguins' world. To me, that all makes sense. Call frequency is influenced by body size, with the largest males being able to produce the lowest frequencies. Such males probably also have the largest fat reserves. If staying power (the ability of an incubating male to continue fasting and not desert the eggs) is a good predictor of breeding success, as I had found during my first stint studying the penguins in 1977, then a big fat male, one able to produce a deep voice, would make perfect sense as the sexiest thing a virgin female could choose.

But what about experienced females? In almost all seabirds, there is an advantage of re-pairing with a previous partner, especially one with which a bird has previously been successful.

We observe that these experienced females, the ones that we had followed from the generator box throughout the previous breeding season, are not so focused on the calling concerts put on by the unattached males. The thing they seem most intent on getting close to is their previous nest site. They go back to it. If their previous partner is there by himself, they will scream their welcoming, which is quickly and simultaneously returned by their old mate, and, this time, is fittingly called a Mutual Call. However, if he is not there, then the urgency to breed in the short Antarctic summer takes over pretty quickly and the females will pair up a new male within a few hours, sometimes even a few minutes.

There are several characteristics of the newly chosen males that we thought might be crucial to the females' decisions. And make no mistake: any decision making or discretion is exercised solely by the females. The males will bonk anything that comes within a flipper's length of them and lays upon the ground. As nests on the outside of the subcolony are most prone to skua predation, we anticipate that females might show a preference for males with central nests. Males that are experienced, especially those that have been previously successful at breeding, we suspect might also prove attractive to the females.

Throughout that courtship period in 1985, we are able to record the identities of all the unpaired males in the subcolony at the exact time each female arrives, and hence match the characteristics of the chosen male against the characteristics of the available males.

Surprisingly, the one feature that stands out as being important to the female is the proximity of the available male's nest to her previous nest site, with females preferring the nearest available neighbor. This makes sense when we observe the subsequent mate switching patterns in the subcolony. If a female's previous partner eventually turns up, she will typically dump her new partner and move back to their old nest site with her old mate (especially if not too many days have elapsed). Similarly, if the previous female partner of the new male shows up and finds him newly betrothed to their neighbor of the previous season, the old partner will fight like hell with the interloper, typically driving her off and leading, again, to another mate switching.

Much of the fighting that had previously been observed in Adelie penguin colonies during the courtship period had erroneously been recorded as males fighting for females—as is the case in many vertebrate species—when in fact, it is mainly females fighting for males.

During that summer of '85, I find two other features that also influence the divorce rate. If the pair had previously reared chicks together successfully, they have a higher likelihood of re-pairing; however, if they are asynchronous in returning to the colony from their long winter migration, then they have a higher likelihood of divorce. In the short Antarctic summer, no Adelie female can afford to wait too long for a previous male partner no matter how good he may have been.

In 1909, Murray Levick has yet to prove how good he is as a partner, but he is starting to show how good he could be to an expedition where, like incubating eggs, endurance might be the difference between success and failure. He is stationed in the coastal town of Shotley on the Suffolk coast east of London. It is here that the Royal Navy operates a training ship as part of a shore-based training establishment, which is used primarily to attract boys into the navy. It is called the HMS *Ganges*, and Levick has been posted here as a medical officer. It is now that he starts to take a special interest in the physical training of others.

Shotley proves to be a small town with a big history—one that has often involved fighting of one sort or another. First the Vikings, then William the Conqueror (it is featured in his Domesday Book) and his bloody sons, then the Hundred Years' War and, finally, two world wars. What is most obvious to me, however, is that as hard as those times might have been, its best days are behind it. The HMS *Ganges* was used to train boys and men for the navy for much of the 20th century but it was closed down in 1976. The mast of the *Ganges*, in a state of disrepair, still hovers on the hill above the little harbor.

Nevertheless, Levick's time in Shotley, developing his interest in physical training, is an important step in moving him closer to Antarctica.

CHAPTER EIGHT

DECEPTION

I t is September 2, 1909. Roald Amundsen is preparing his expedition
to become the first person to get to the North Pole. His childhood
hero and erstwhile mentor, Fridtjof Nansen, has given him the use of
the *Fram*, the ship specially built for polar conditions.

Yet, on this day, Amundsen hears astonishing news from another
of his mentors. Frederick Cook, the doctor and his savior from the
Belgica Expedition, sends a cable from the Shetland Islands in northern
Scotland to say that he has reached the North Pole. He claims to have
gotten to the pole on April 21, 1908, but, because the journey back
has been so long and hard, it has taken him eighteen months to get to
anywhere civilized enough to be able to send a telegram.

Incredibly, within a week, Cook's own mentor, Robert Peary—with
whom Cook had crossed Greenland and from whom he had learned
the art of polar travel—claims to have gotten there too. In a separate
expedition, Peary says that he arrived at the North Pole on April 6,
1909. Peary, who is endowed with a spectacular mustache that looks
like a couple of bull walruses facing off on either side of his lips, now
turns his formidable bite on his former protégé. He and his supporters

immediately set about discrediting Cook's character and claim. Particularly damning to Cook's claim of priority, they say, was his announcement three years earlier that he had been the first to summit Mount McKinley (renamed Denali) in Alaska. This seems highly unlikely, given that the photo—which Cook said was taken from its peak—is demonstrably taken from a nearby mountain.

Fast on the heels of Cook and Peary's announcements, Robert Falcon Scott makes one of his own. On September 13, 1909, the same day that Kathleen gives birth to their son, Peter, Scott publicly announces his intention to mount another expedition to Antarctica to "reach the South Pole and secure for the British Empire the honor of that achievement."

Murray Levick, along with eight thousand others, puts his name forward, and undoubtedly his prior acquaintance with Scott helps get him the job as surgeon and part-time zoologist on the expedition. Additionally, Levick has already made a name for himself as a doctor in areas that make him a highly suitable appointment for the expedition: diet, physical fitness, and training.

At this stage, Levick has never seen a penguin, never had any inclination to study them, nor, indeed, to do any zoology. He is signing on because, like me sixty-seven years later, he simply wants to go to Antarctica.

It is hard to say where that desire in Levick has come from but when he joined the Royal Navy and went through his initial naval medical training, one of his classmates was Alister Mackay. Dr. Mackay became the surgeon on Shackleton's Nimrod Expedition. It is easy to imagine that Levick has discussed the trip to Antarctica with his classmate, or at the very least is aware of it, and that he longs for such adventure himself.

Others, however, are interested not so much in the adventure but in securing the achievement of reaching the South Pole for themselves and their own countries. The *New York Times* reports, three days after Scott's public announcement, that Robert Peary is now seeking to be the first person to get to the South Pole, thereby securing for the United States the honor of being the first to get to both ends of the Earth. Similar intentions are announced by other would-be explorers: Wilhelm Filchner in Germany, Jean-Baptiste Charcot in France, and Nobu

Shirase in Japan. There is, however, no such announcement coming out of Norway: Amundsen maintains that the achievement of his friend Frederick Cook in getting to the North Pole—Amundsen remains loyal to the man he views as saving his life when on the *Belgica* and continues to support Cook's claim—does not affect his forthcoming expedition to the North Pole, as, he maintains, its primary focus is "oceanographic investigation" rather than being first to the North Pole.

Raising money for polar expeditions—which requires ship support, supplies, and men to be paid for several years—is no easy matter, even for people with the reputations of Amundsen, Peary, Scott, and Shackleton; even with the backing of august bodies like the Royal Geographical Society. The men count on moneys received from public talks as a way to help balance their books. It is why Scott gave his lecture at Whitworth Hall, where he drew a penguin on the blackboard, and it is why Shackleton is looking to follow suit.

Shackleton's finances are particularly strained. Men like Mawson have remained unpaid for their two years of service, even months after the expedition is over. Accompanied by Emily, Shackleton embarks on an extensive public lecture tour that takes him to the United States and various parts of Europe. During a stop in Oslo in October 1909, Shackleton talks to a hall full of attentive members of the Norwegian public. In the audience there is one man who is more attentive than the rest: he is tall, quiet, and has a large nose, but it is his eyes that most captivate Emily. When her husband speaks, she notes of the attentive audience member, "a mystic look softened his eyes, the look of a man who saw a vision."

Roald Amundsen absorbs every word from Shackleton. How the access to the Ross Ice Shelf—discovered initially by his childhood friend Carsten Borchgrevink and visited by Scott and Shackleton in 1902, at the nearby Balloon Bight—has been altered by the calving Ross Ice Shelf and become a large bay, the Bay of Whales. How the view and photographs from Scott and Shackleton's near-disastrous ballooning attempt had provided a view of a flat ice shelf extending as far as the eye could see, a wide, white highway stretching, it seemed, almost to the South Pole.

Although Amundsen does not admit it to anyone but himself, by the time he hears Shackleton talk, he has already concluded that he should head south and not north. That is where the remaining honors lie, the one remaining prize.

It is December 14, 1909, and Shackleton is at Buckingham Palace to collect the ultimate honor for just getting close to the South Pole. Wearing tails and a sword, Shackleton is knighted by King Edward VII, the man whose photo graces the wall of the hut at Cape Royds. The king touches a kneeling Shackleton on the shoulder with his own sword, making a connection between the two charmers—the two philandering charmers. Sir Ernest Shackleton. A knighthood still alludes Scott, and when simply saying the words "Sir Ernest," they must have caught in his throat no less than if Shackleton and the king had both shoved their swords down his throat at the same time.

Meanwhile, the controversy over the competing claims of Cook and Peary has continued to rage. A week after the ceremonies at Buckingham Palace, when in addition to Shackleton being knighted, all members of the Nimrod Expedition had been awarded the Polar Medal by the king, a commission at the University of Copenhagen declares that there is insufficient evidence to support Frederick Cook's assertion of having gotten to the North Pole. The detailed navigational records that might have substantiated Cook's claim had been left in Greenland with a hunter, Harry Whitney, who was supposed to take them back to America for Cook. Ironically, Whitney ended up returning to the States on Peary's ship, and it had been Peary himself who had refused to transport Cook's diaries and records on his ship. Left behind in Greenland, the records were never seen again. Curiously, however, Peary does not allow his own records to be scrutinized. These are held by the National Geographic Society and, in accordance with the wishes of the Society and Peary's family, remain locked away from all who might want to see them.

Irrespective of the bickering about Cook, to whom Amundsen remains loyal and, at least outwardly, professes to believe, he is forced to accept that even if Cook did not get to the North Pole, then surely Peary did. That prize has been won.

Secretly, he sets about planning to go to the South Pole instead.

—◈—

It is August 22, 1988, and the *New York Times*, which had worldwide exclusive rights to Peary's story at the time he claimed to have gotten to the North Pole, as good as publishes a retraction. *National Geographic*, which celebrated its centenary that year, finally opens up the Peary archives and publishes a review of Peary's data by none other than Sir Wally Herbert, who, in the same year that Neil Armstrong was stepping onto the moon, became the first undisputed person to travel to the North Pole by dogsled. Peary's records prove sketchy. The pages of his diary where he should have recorded his positions are blank on the days that he was supposed to be at the Pole. Instead, a loose page has been inserted merely saying, "The Pole at last!!!" His claims about the distances he traveled per day—over seventy miles—seem highly improbable, and also, when he did calculate his position elsewhere in the diary, he did not correct for the effect of currents and the movement of the ice over which they were traveling: all of which are important to accurately determine one's location in the Arctic. Herbert concludes that Peary was likely thirty to sixty miles away from the North Pole when he claimed to have been there.

I am struck by the deep irony of all this.

—◈—

In 1909, the competing claims by Cook and Peary that they have reached the North Pole really do affect Roald Amundsen, despite his public protestations to the contrary. They transform him into an otherwise unwitting competitor with Scott, in what will now become an out-and-out race to get to the South Pole. And that, as much as anything, will be responsible for Murray Levick becoming a penguin biologist.

Douglas Mawson also longs to go south to the Antarctic. But not for any glory attached to getting to any poles, magnetic or otherwise:

he wants to go for the science, for the geology, for what the big white continent can tell us about the history of the world, and in particular the relationship between the Antarctic continent and Australia.

Mawson was one of the few men from the Nimrod Expedition who did not attend the ceremony at Buckingham Palace when Shackleton kneeled before the king as Ernest and rose again as Sir Ernest. At the same ceremony, King Edward VII had presented all the other members of the Nimrod Expedition with the Polar Medal, giving Mawson's in absentia.

Mawson does, however, arrive in London a month later, where he arranges a meeting with Scott in mid-January 1910. Scott offers to take him as part of his new expedition to Antarctica, the Terra Nova Expedition. Mawson is insistent that he will go only if he can be dropped at Cape Adare with a party of three other men so that they may explore the coastline west of Cape Adare, which he believes will reveal evidence of it having been once attached to Australia. Scott is reluctant to add another base to his expedition when he has already committed to setting up his main base in McMurdo Sound and a secondary base on Edward VII Land. Mawson says that, in that case, he will mount his own expedition to Cape Adare. A second meeting takes place between the two men, with Bill Wilson sitting in, but again, makes no headway.

It is January 26, 1910. Scott invites Mawson for dinner to continue their discussions. There Mawson meets Kathleen and is immediately drawn into an orbit around her, along with a whole solar system of admirers of her celestial body that include George Bernard Shaw, James Barrie, and Auguste Rodin.

Unable to persuade Scott, Mawson turns down a place on the Terra Nova Expedition and looks for ways to organize his own expedition to Cape Adare. Sir Ernest Shackleton is continuing his speaking tour in Europe. When Mawson meets Shackleton upon his return, the new knight agrees to help, but on the condition that it becomes *his* next expedition to Antarctica with Mawson as the chief scientist.

Shackleton, anxious not to be seen as treading on Scott's toes or territory in any way again, writes to Scott:

I am preparing a purely Scientific Expedition to operate along the coast of Antarctica commencing in 1911. The Easterly base is Cape Adare. . .

Scott, meanwhile, continues to assemble his expedition members and the equipment they will need if they are to have a chance of making it to the South Pole. One of the innovations that Scott is banking on is the use of motor sledges, which are, in effect, newly invented motor cars with tracks. In early March, he travels with Kathleen to Norway to see tests of the motor sledge, which promptly, and disconcertingly, breaks an axle.

While in Norway, they call upon Fridtjof Nansen, who though he doesn't particularly like Scott, he, like Mawson and so many others, is drawn to Kathleen. Nansen is appalled to learn that Scott intends to rely chiefly on unproven gasoline-powered motor sledges and ponies rather than Eivind Astrup's proven means of polar travel: dogsleds and skis.

Not everyone, however, is oblivious to the example set by Astrup. At about that moment, on the other side of the world, Bertram Army-tage, who had been part of Shackleton's Western Party with Raymond Priestley, checks into a room at Melbourne's exclusive Melbourne Club. He puts on a tuxedo and the silver Polar Medal he had received from King Edward VII at Buckingham Palace just three months earlier. He places a towel on the floor. Then he lies upon the towel, places a pistol to his head, and shoots himself.

Nansen regards Scott's own plans as, if not suicidal, then stupid. He recommends strongly that Scott at least take a young Norwegian skiing expert with him, Tryggve Gran, so that he might teach Scott's men how to ski properly.

Gran has met Amundsen before, and given that Amundsen has announced plans to go to the North Pole, Gran offers to help arrange a meeting between Scott and Amundsen so that they may consider coordinating the scientific measurements they shall be taking at opposite ends of the Earth. Gran sets up the meeting through Amundsen's brother Gustav. However, when Scott and Gran arrive at Amundsen's home on the banks of Bunnefjorden, they are met by only Gustav. They wait for an hour, but Amundsen, although aware of the meeting, curiously does not show up. This is understandable perhaps, if not forgivable, given what he is really up to.

Outwardly undeterred by the Cook and Peary claims, Amundsen has been, to all intents and purposes, proceeding with preparations for his expedition to the North Pole. Secretly, however, aided by another brother, Leon, Amundsen is making plans to go, instead, to the South Pole. He keeps this all a secret from the crew he is assembling for fear of losing the backing of Nansen and the new king of Norway, King Haakon VII, who are supporting the expedition on the basis that it will increase Norway's prestige in the Arctic regions. The Norwegian government, with Nansen's blessing, has offered Amundsen Nansen's ship the *Fram* for the expedition. Nansen, however, does not want Amundsen to take just his ship: he presses Amundsen to take with him Hjalmar Johansen, the man with whom he had tried, gallantly, if vainly, to reach the North Pole. Johansen is vastly experienced, an undeniable expert in polar travel, but also a drunkard whom Amundsen foresees is likely to cause trouble. Amundsen is extremely reluctant but he relents because he does not want to jeopardize, in any way, the support from Nansen or the expedition's sponsors.

Amundsen's house and property, which he calls Uranienborg, is sited a few yards from the shore of the Bunnefjorden, south of Kristiania. It is pretty much as far from the water as this man of the sea is prepared to be. Yet there is just enough room between the front steps and the pier to build the hut he plans to take to the Antarctic and set up on the ice shelf discovered by James Clark Ross at a place first visited by his childhood playmate Carsten Borchgrevink and now called the Bay of Whales by Sir Ernest Shackleton. He is highly secretive: to test the hut, he lives in it, but no one else is allowed near it. No one, that is, except for Sigrid Castberg. There they make love on the beds of the hut among the reindeer skins. These are the happiest moments of Amundsen's life.

It might be supposed that this recluse, who has, until now, barely known the touch of a woman and certainly not the gut-wrenching euphoria that love can bring, deliberately chose a married woman to be his partner in that it lessens the complications, making it easier to move on when the ice calls loudest. That is at least as far from the truth as Peary had been from the North Pole. Amundsen implores Sigrid, his first true love, to leave her husband and marry him. She refuses.

It is only then that he acts like the coldhearted explorer that he is; the man who is capable, with dispassionate ease, of weighing risks and the value of lives; the man who can kill and then eat with relish the dogs who have served him so well. He cuts her off completely.

It is May 6, 1910. The monarchy and its approach to sexuality is about to change again to more puritanical ways, that is, if you can turn a blind eye to a little bit of incest. Edward VII (the Caresser) is able to caress no more. As heavy a consumer of cigarettes and cigars as he is of women, he has a heart attack and dies after a period of illness that is no doubt not helped by his smoking, if not his whoring too. Remarkably, Alice Keppel, his last great mistress, is permitted by Queen Alexandra see him on his deathbed. He is succeeded as king by his second son, King George V.

It is true that, prior to becoming king, George V had tried to marry his cousin, and after that was quashed by his mother and aunt, he married another, albeit somewhat more distant, cousin, Princess Victoria Mary (known as May), who had earlier been engaged to his older brother. It had been the death of his brother from pneumonia that had left George next in line to the throne.

In the dozen years following their marriage in 1893, George and Mary had six children and gave every indication of being totally devoted to each other, while his father continued to hang out with princesses and prostitutes, right up until his death..

In the peculiarly inbred partnerships that characterize European royalty, which makes studying the mating behavior of penguins seem straightforward by comparison, by the time King George V has his coronation, another of his cousins has become Norway's first king, King Haakon VII, and is married to yet another cousin, Queen Maud.

It is June 1, 1910, and the *Terra Nova* is leaving behind England and its new king and queen. On board are some sixty men. Among the officers is Murray Levick, who has been released from his naval duties at Scott's request. Scott has gone out of his way to promote the expedition as one intent on new scientific discoveries first and an attempt to get to the South Pole only second. While that is disingenuous to say the least—because the bulk of the expedition's equipment, animals, and human cargo will be in the Antarctic to support the push for the pole—he holds up his plan for the Eastern Party as evidence for this.

It is to be led by Lieutenant Victor Campbell, the thirty-four-year-old first officer of the *Terra Nova* and one of the few men, like Scott, with a child. Campbell has an eight-year-old son, although his marriage is strained following his wife's depression after the death of her sister, and secretly he is glad for the opportunity to spend time away exploring new lands. The Eastern Party is to explore the land named by Scott in 1902 after the now dead king: King Edward VII Land. This stretch of land, visible at the eastern edge of the Ross Ice Shelf, remains unexplored either through Shackleton's failure to get there, or as Scott believes, Shackleton's betrayal.

Sir Clements Markham, Scott's longtime supporter, is at the wharf to see them off. The *Terra Nova* makes its way along the southern coast of Britain, stopping at various points until it makes its last port of call in Cardiff. It is June 15 when it finally pulls out of Cardiff and heads into the Atlantic. As Campbell records in his diary that day, in typical perfunctory fashion:

> *We left Cardiff weather fine and calm. Several pleasure steamers came some way with us, the tug taking off Captain and Mrs. Scott and friends.*

Scott will stay behind to continue necessary fundraising and shall follow six weeks later in a fast boat in order to meet the *Terra Nova* in South Africa. Leaving their son Peter behind, Kathleen shall travel with him, together with the wives of Bill Wilson and the second in command, Teddy Evans.

The *Terra Nova* sets sail for Madeira, the archipelago of subtropical islands in the Atlantic belonging to Portugal. They arrive on June 23 and stay there for three days.

It is September 3, 1910. The *Fram* pulls into Madeira too, where it shall stay, also for three days, taking on fresh water and fresh food while also mending a broken propeller. The *Fram* had earlier left Oslo on June 3 and traveled down to Amundsen's home on the Bunnefjorden. There the hut, which Amundsen had built on his lawn and tested so thoroughly with Sigrid, was dismantled, each part numbered, and then stored on the ship. At midnight, at the start of Norway's Independence Day on June 7, the *Fram* snuck away quietly with no fanfare at all. As the *Fram* sailed past Kristiania, across the fjord, in his office at the top of the redbrick tower of his home, Polhøgda, Nansen watched the ship that had served him so well slip out of the fjord toward the open sea under the command of Amundsen. He would call it "the bitterest moment in my life."

They had spent a month in the northern Irish Sea, ostensibly on a brief oceanographic survey, but what Amundsen was really doing was testing out his men, the ship's engine, and their procedures. They went back to Norway, made some adjustments to their fuel, loaded supplies, and ninety-seven dogs from North Greenland that had been specially chosen for their strength. They had finally left the Norwegian coast behind on August 10, ostensibly bound for the North Pole. Their route supposedly is taking them south around Cape Horn in order to get to the Pacific Ocean and the Bering Strait that is to be their gateway to the Arctic.

Just three hours before their departure from Madeira, Amundsen calls all the men on deck for a meeting. He stands before them with a large map of Antarctica pinned to the mainmast. It is then that he confesses that he has deceived them: they are not going to the North Pole but to the South Pole. The only ones privy to his secret beforehand are his brother Leon, the first mate Kristian Prestrud, and Lieutenant Fredrik Gjertsen, the young officer who had already been entrusted with carrying the train on Queen Maud's dress during King Haakon VII's coronation in 1906.

Amundsen explains the reason for his deception, saying that had the secret gotten out while still in Norway, they would have most likely lost the support of their sponsors. The men are stunned. There is stony silence. One by one, Amundsen asks them in turn if they will come with him. Those who leave the expedition now will have their passage paid back to Norway. Yet there is no need for such a contingency: to a person, they all say "yes" to their leader.

Leon Amundsen disembarks just before the *Fram* sails out of Madeira. He carries with him letters written by Roald Amundsen to the king and Nansen, which three weeks later he will get to both men through an intermediary, having given the *Fram* enough time to get well on its way to Antarctica. He also carries with him letters written by the men to their families explaining their new circumstances that he will post and, then, afterward, cable Scott and tell the press.

Amundsen's letter to his childhood hero ends with what sounds eerily similar to the way Jesus may have expressed himself for having deceived Nansen so deliberately and completely:

> *I beg your forgiveness for what I have done. May my coming work help to atone for that in which I have offended.*

Forgive me father, for I know not what I have done. Though ye believe not me, believe the works. Indeed.

Among the first to encounter Amundsen and his "works," will be the Eastern Party. And Murray Levick is about to become part of that.

CHAPTER NINE

THE EASTERN PARTY

The *Terra Nova*, with Scott in charge, is *en route* between South Africa and Australia. Still needing more financial support for his expedition, Scott has sent Bill Wilson and the wives ahead in a fast ship so that the good doctor may continue fundraising there until they can all meet up in Melbourne.

It is during this period that Scott and Campbell start to divvy up who shall be in the Eastern Party and who shall be in the Shore Party to support the push for the Pole. Campbell is to lead the Eastern Party and, as they shall be on their own for upward of a year, he needs a medic. There are two surgeons assigned to the expedition: Murray Levick and Edward Atkinson, who is five years younger than Levick. They decide that Levick is the obvious choice for the Eastern Party, given his expertise in diet and physical exercise—although, presumably, such skills would also be valuable on the proposed march to the pole.

In truth, Scott is just as happy not to have Levick in his lot as he is for Campbell to have him. Despite having picked Levick for the expedition, Scott is none too impressed with him. As he puts it:

He seems quite incapable of learning anything fresh. Left alone, I verily believe he would do nothing from sheer lack of initiative.

In fact, the Eastern Party will require Levick to do triple duty: in addition to being its medic, he is expected to act as the Eastern Party's zoologist and photographer.

As for a geologist to map and describe the hitherto unexplored Kind Edward VII Land, that job is to go to the Australian, Raymond Priestley, who had been with Shackleton on the Nimrod Expedition. Priestley is in Sydney and shall travel independently, meeting the *Terra Nova* when it gets to Lyttelton in New Zealand.

When it comes to deciding upon the three able-bodied seamen to be assigned to the party, Campbell has already taken a shine to the handsome and physically impressive petty officer, George Abbott, the tallest man on the expedition. Very fit and already at thirty with graying hair, Abbott has been learning taxidermy from Dr. Bill Wilson, who describes him as "an exceedingly nice gentlemanly fellow and a tower of strength." It is almost inevitable, then, that Abbott should be known as "Tiny" among the men. Petty Officer Frank Browning is a torpedo expert who grew up on a farm and, at twenty-eight, is selected because he is very adaptable. Short, lithe, and dark-haired, he goes by the even more unlikely name of "Rings." The last of the three is the young Able Seaman Harry Dickason, who, at twenty-five, is already proving to be a skilled cook. Almost as short as Browning, the light-haired Dickason has the most obvious of nicknames: "Dick."

An investigation into how Murray Levick could go from being picked for the Eastern Party to becoming a man fascinated by penguin sex would ideally involve interrogating all the members of the Eastern Party: Levick, Campbell, Priestley, Tiny, Rings, and Dick. Except for one problem: they are all long dead.

I am in Cambridge, England's beautiful university town with its cobbled streets and colleges, pubs and punts, all of which have been traversed by some of civilization's greatest minds. From Isaac Newton to Ernest Rutherford to Stephen Hawking, I cannot help wonder which of them may have put their boots on the same cobblestones where I place my feet now. I am there to go to the University of Cambridge's Scott Polar Research Institute. It was founded in 1920 by two members of the Terra Nova Expedition, Raymond Priestley and Frank Debenham, to be used primarily as a repository for materials from polar exploration. Its current location is a rather bland, gray building on Lensfield Road. There is the usual security to get in, though perhaps not as tight as that at Tring. You do need to book ahead to reserve one of the few desk spaces allocated for research in the tiny reading room. I have booked a whole week. Eventually, I am led into a small gray room and ushered to a desk. There are already two others sitting at theirs, eyes down, reading. I am shown the blue catalogue files and then told that I can ask to see literally anything in their collection. Anything. Letters written by Scott. Letters written by Shackleton. Letters written by Kathleen Scott. Anything at all.

What I am most interested in are the diaries of the men who were part of the Eastern Party. Those of the leader, Victor Campbell, are not held here, but pretty much everything else is. Campbell's diary, however, has been published as a book. Indeed, Priestley's account of the expedition, based upon his diaries, had been published in 1915. And, nearly a century later, the terse diary of Harry Dickason with its brief entries had been published. I am particularly interested in the diaries of the much more loquacious G. Murray Levick, only one of which is available in published form. I ask for one and it is brought to me and placed on what is a sort of soft beanbag for support. My fingers shake with excitement as I reach for it.

I am able to smell Levick on the pages, touch him, climb inside his head. These are not zoological notes, his writings about science. These are the thoughts of the man born George Murray Levick. What moved him, what didn't. What happened to him, yes, but, most intriguingly, what he wished would happen to him. What he thought of his fellows and what he thought about life.

It feels like such a privilege to be here. At last it seems that I have caught up with my man, my Amundsen, and while I cannot talk to him directly over the next week, this is surely going to be the next best thing. I barely notice the comings and goings of the others in the small room.

⸺∘∘∘⸺

It is October 12, 1910, when the *Terra Nova* arrives in Melbourne's Port Phillip Bay. It is a huge wide bay and Kathleen insists on Wilson and the wives going out to meet the ship in a launch. Kathleen has been unfavorably disposed to both Hilda, Teddy Evans's wife, and Oriana, Bill Wilson's wife. This has led to tensions between the three of them, causing Wilson to reveal that he is ill-suited to be a penguin:

> *I hope it will never fall to my lot to have more than one wife at a time to look after.*

Scott and Kathleen, along with the other two couples, then take the launch back to the city and the hotel where they are staying. At the hotel, Kathleen hands Scott the mail that is waiting for him, including an envelope containing a cable.

Scott opens it. It contains nine words. Nine words that will forevermore change his life and, indeed, that of Levick too:

BEG LEAVE TO INFORM YOU FRAM PROCEEDING ANTARCTIC AMUNDSEN

This time it is Scott's turn to be secretive with his crew: he does not tell them for fear of affecting morale, though he does ask Gran what he can make of the cable. Gran is no help, and so Scott cables Nansen asking if he can tell him Amundsen's destination. Nansen's reply is even more concise than Amundsen's cable had been: UNKNOWN.

It is October 28, 1910. The *Terra Nova* pulls into Lyttelton Harbor in New Zealand. In the middle of the harbor is a small green island, Quail Island, which is being used as a quarantine station for livestock and any arriving immigrants to New Zealand who are deemed sick. From 1906, it has also housed a small leper colony. For the last six weeks it has been home to forty-nine exotic animals: nineteen white Manchurian ponies and thirty Siberian dogs, which have been brought there by Cecil Meares and Wilfred Bruce, one of Kathleen's older brothers. Scott has determined, based upon Shackleton's experience on the Nimrod Expedition, that ponies and motorized sledges are his best means to get to the pole. Fridtjof Nansen had begged Scott to take dogs and, in deference to him, they are taking thirty, though that is not nearly enough needed to get to the pole, for they are intended only for support work: assistance with setting up food depots and the like.

Scott may have viewed Levick as lacking in wit and initiative, but Levick shows plenty of both on the journey down to Antarctica as he squirrels away supplies and equipment that might be useful to the Eastern Party, knowing full well that Scott's own Polar Party will have first call on such things once they are in the Antarctic.

The *Terra Nova's* last port of call is Port Chalmers, my home town. On a hill above my house sits a monument to Scott and his crew: a thirty-foot-high cairn of local stones with an anchor atop. I go past that monument every day and it never fails to connect me to Levick and those times. As I look down on our small town, with most of its houses and shops unaltered since Scott's time, it always makes me smile to imagine that as much of the crew was out enjoying the last delights of civilization, be they of the alcohol or feminine kind, Levick was sitting in his cabin stuffing goodies under the mattress. Levick alludes to the feminine attractions of Port Chalmers when in the first entry in his Antarctic diary he writes:

> *I think most of us feel regrets a (sic) leaving New Zealand, as we have all made friends, and some of us I dare say, more than friends.*

The Otago Harbor is beautiful, a narrow band of sheltered and shallow water that draws the eye up to Taiaroa Head with its lighthouse and the narrow entrance to the harbor slotted between the headland and a spit of sand on the other side. The departure of the *Terra Nova* from Port Chalmers in the mid-afternoon on November 29, 1910, is notable less for its cheering send-off than it is for the almighty row that takes place in the hotel between Kathleen Scott and Hilda Evans just before the ship leaves, which becomes a three-way battle when Oriana Wilson steps in to try to break them up.

Titus Oates, the man in charge of the ponies, says in a letter to his mother:

> . . .*there was more blood and hair flying about the hotel than you would see in a Chicago slaughter house in a month.*

The women accompany their husbands on the ship as far as Taiaroa Head. When it comes time to transfer to the tug, Kathleen, unlike the other wives, chooses not to kiss her husband goodbye. She will say later that she did not wish to make him sad in front of the other men. Yet this stiff and formal parting speaks volumes about their pairing.

In penguins, reinforcement of the pair-bond, which they do by mutual calling, is a good predictor of whether a pair will stay together or divorce.

My finding from the summer of '85 that pairs that had been unsuccessful in their breeding attempt the previous season have a higher likelihood of divorce can be explained precisely by the lack of reinforcement of their pair-bond. When male and female Adelie partners greet each other, they do so by engaging in a Mutual Call. The pair trumpet loudly in unison, standing breast to breast, their bills pointed skyward, waving their heads and necks about each other.

It occurs particularly when a bird that has been at sea arrives at the nest to greet its partner. It is the way they recognize each other and affirm their bond. In humans, we kiss and hug each other.

Because the pair takes turns going away to sea for about two weeks or more during the first two incubation stints, if they lose their eggs for whatever reason, they will have only two or three occasions to reinforce their bond: during courtship and at the nest changeovers that occur at the end of the first and second incubation spells. After that, the birds change places on the nest virtually every day or two, so there are many, many more times to perform their mutual greeting and reinforce their pair-bond for those pairs that manage to raise at least one of their chicks to fledging. During the Antarctic winter, the pair are not together as they migrate north. When they return to the colony to breed at the start of the following season, pairs that have had the opportunity to frequently reinforce their bond are more likely to reunite. Pairs develop a stronger relationship or bond borne of their frequent mutual calling at nest relief, which is something that only successful pairs experience.

Kathleen boards the tug beneath the white and red lighthouse of Taiaroa Head. She says goodbye to her husband, Captain Robert Falcon Scott, but deliberately refrains from kissing him. Deliberately refrains from reinforcing their pair-bond.

She will never see him again.

I am leaving New Zealand on a Russian ship, the *Akademik Shokalskiy*, bound for the Ross Sea, as I attempt to follow in Levick's footsteps. I have the potential to get seasick just looking at water and I have purchased about $300 worth of various seasick tablets.

All starts calmly enough, but within a day we are battered by a massive storm. Waves break over the bow of the ship as it plunges from one huge wave to the next. In my bunk, I am thrown with considerable force from one end of the bulkhead to the other with each pitch and roll of the ship. I take several different types of pills to no avail: I am seasick, but, then, so is almost everyone else, including most of the crew and the onboard doctor. After two days, the wind abates and we are left to count the damage. Most of us are badly bruised from being thrown about. One person has a broken collar bone. Another has internal bleeding and needs to be airlifted off the ship in a dramatic long-distance helicopter rescue when we reach the Auckland Islands.

It is November 29, 1910. The *Terra Nova* has turned south after clearing the heads at Otago Harbor. It is supremely overloaded. Below decks is completely full and they have been forced to store the three motor sleds, forty tons of coal, two thousand gallons of gasoline, and the pony fodder on the upper deck. It is not much bigger than the *Fram*, yet the *Terra Nova* has sixty-five men, nineteen ponies, thirty dogs, and three motor sleds on board. By contrast, *Fram* has just nineteen men aboard her, each with his own small room, and ninety-seven dogs. The overladen *Terra Nova* is slow and wallows in the sea, causing many of the men to be seasick. Campbell notes that, "We must hope for fine passage," but that is not to be. Three days out from New Zealand, they are hit by a frightful storm.

The ship is leaking and begins taking in lots of water. Their pumps fail. The men are forced to bail her out with buckets and a hand pump, which as Campbell nonchalantly observes is, "very slow work as the men were constantly being washed off their legs." Throughout the raging storm, they scramble frantically, trying to clear the pumps while being thrown about and throwing up.

By the time the storm ends and they can assess their damage, two ponies are dead, and a dog and ten sacks of coal have been washed overboard.

It is Christmas Day 1910. The progress of the *Terra Nova* has been inhibited by dense pack ice. To date, the men's only real interaction with the wildlife has been to kill it and eat it. Crabeater seals are served as steaks, which according to Levick, "is excellent. More tender than beef steak and quite as good to eat." They stew Adelie penguins, which Levick finds:

> *. . . a really first class bird—rather like blackcock to taste, but a good deal better—the flesh is black, like seal meat.*

His appreciation of seal and penguin meat is probably just as well, given the diet that awaits him in Antarctica. In fact, the penguin meat is so good that stewed penguin forms part of their first Christmas dinner in the ice.

The day before, on Christmas Eve, Levick noted the men sang, "in a horrible discordant manner," to Adelie penguins that had gathered about the stationary ship. At the end of the performance, the penguins stood around, "cawing and bowing their appreciation."

Eventually the pack ice eases and the *Terra Nova* is able to make its way down to Ross Island. The sea ice has not broken out completely, and they cannot get to Scott's preferred landing at Hut Point, his old site from the *Discovery* days. Instead, they decide to set up the hut and base a little farther north at Cape Evans, where they arrive on January 5, 1911. They begin the long and weary process of unloading. They soon learn that this is an environment where the line that separates success and survival from failure and fatality is a very fine one, indeed.

Herbert Ponting, the expedition's photographer, or "camera artist," as he prefers to call himself, tries to photograph a pod of six killer whales that are attempting to hunt penguins at the edge of the sea ice. He takes his camera and tripod to the very edge of the sea ice but the whales, seeing him, go under the ice, coming up and breaking it into small floes. Ponting is left rocking on one of the floes when one of the whales rears out of the water, its head over the edge of the floe, trying to grab him. Ponting jumps from floe to floe and is able to get back safely to the fast sea ice and, as Levick admiringly notes, "To his great credit he saved his camera and tripod."

Levick is a great admirer of Ponting and desires desperately, given his new duties for the Eastern Party, to learn the craft of photography from Ponting. However, Ponting is less enamored with the persistent Levick, and to Levick's disappointment, largely ignores him. Levick records in his diary:

> *I find I can't get any information out of Ponting—He won't give anything away as to his methods of exposure, developing, etc, though I should not think he can lose much by teaching me.*

A second calamity is potentially more serious as far as the expedition's aims go. The men have unloaded the third of the three motor sledges from the ship and are pulling it across the sea ice toward the land where the hut is being erected, when half a mile from the ship, the ice suddenly gives way and the heavy motor sledge sinks, immediately pulling two men with it into the water. One of them is Priestley and he is by far the one in the most precarious position: pulled completely under and, at one stage, under an ice floe. It is my childhood inspiration, Apsley Cherry-Garrard, who comes to the rescue. He skis over to the men left standing on the broken ice—who have by now hauled Priestley and the other seaman out of the freezing water—with a lifeline that is used to haul each of them to safety. There is no such happy ending for the motor sled, however: it is in 120 fathoms of water and well beyond being hauled to safety.

It is January 16, 1911. Levick, Campbell, and Priestley leave the main party to get their digs sorted out at Cape Evans and ski over to Cape Royds; the first persons to go there since James Murray shut the door on the hut and the opportunity to be the world's first penguin biologist.

Scott himself has little desire to go there because of it being Shackleton's base. As he writes in a letter to Kathleen, "Always I have had the feeling that Cape Royds has been permanently vulgarized." It is the

reason why he had initially considered setting up his base at Cape Crozier. "There is no trail of Shackleton there," he says to her, revealing how
deeply he is tormented by the big Irishman—the big knighted Irishman.

Levick and Priestley are first to go to Shackleton's hut, Priestley's old
home from the Nimrod Expedition. They break away ice around the
door and enter the large dark room, as the windows had been covered
with shutters. Levick lights a candle that is by the door. His response
on seeing it is uncannily like mine a lifetime later:

> In the middle of the hut was a long table with the remains of
> their last meal.
>
> A tray of bread scones stood on a box, and tins of every
> description of food stood in piles on shelves round the walls,
> whilst the mens (sic) beds stood at intervals around the sides. All
> the little personal belongings of the late expedition lay about,
> as they had left them on their hurried departure.

It really was like they had just stepped out of the hut. No one felt
that more than Priestley, who found the experience of going back to
the hut where he had lived "very eerie."

> I expect to see people come in through the door after a walk over
> the surrounding hills.

Levick takes a walk through the penguin colony, observing the adult
birds "bringing in food for their little downy youngsters." Apart from
acknowledging the hard work evidenced by these parent birds there is
not a scintilla of enthusiasm for the penguins, nor any evidence that he
has any wish to study them himself. To the contrary, he writes in his
diary that, "Their habits and characteristics have been so well described
by Wilson in his 'Discovery' reports that it is no good repeating them
here . . ."

He is much more focused on killing Weddell seals, which they
cache in the snow and ice for later, and pilfering what they can from

Shackleton's hut for their own ends on the Eastern Party. He is more taken with the hoofprints he can see of Shackleton's ponies that remain in the snowbanks than he is with the penguins.

It is January 26, 1911, and, following a speech from Scott, the *Terra Nova* leaves Cape Evans with all the Eastern Party onboard, intent on establishing their base to explore King Edward VII Land. While en route to Antarctica, Scott had discussed with Campbell a possible change from the original plans. Rather than going right along the Ross Ice Shelf until they get to King Edward VII Land, Scott proposed that perhaps they should land at the Bay of Whales, or Balloon Bight, as he insists on still calling it, and use that as their base to get to King Edward VII Land. This provides two advantages from Scott's perspective. It enables the party to check up on Shackleton's explanation that Balloon Bight and the easy access it afforded to the shelf has disappeared—something that Scott does not quite believe or trust about Shackleton's explanation for abandoning his original commitment to stay away from the McMurdo Sound region, which Scott regards as his territory. Additionally, it will save precious coal, allowing the *Terra Nova* to explore the coastline west of Cape Adare on its way back to New Zealand, where it shall spend the winter. Ironically, the hard-done-by Scott is anticipating exploring the very region that Mawson has proclaimed to be his focus, if not his territory.

The ship stops briefly at Cape Royds where this time Levick's only interest in the penguins is to collect twenty of them for food, along with picking up the frozen seal meat they had cached there. Once they leave behind Ross Island, they travel eastward along the Ross Ice Shelf, keeping close to its sheer face.

<hr />

I am on the *Shokalskiy* doing the same, sailing within yards of the hundred-foot-high sheer cliffs of ice. It is daunting: an impenetrable, perfectly flat-topped block of ice stretching as far as I can see. The occasional crash of falling ice and the scattered icebergs that have calved off its face are the only hints that this giant sheet of ice is alive and moving. The sea is deep, an inky black, yet where its swell undercuts the edge of the wall of ice, the light reflects as a bright turquoise even though the day is dull and cloudy. The walls are impossibly steep and chiseled, like a giant sculptor—a giant Rodin or Kathleen Scott—has hacked at the edge of the shelf with a giant chisel and mallet. A pair of minke whales swim in the narrow space between us and the wall of ice. Quite fitting really, because at the only place where the wall could be breached, Shackleton had found such an abundance of whales that he named it the Bay of Whales.

At first, the *Terra Nova*, captained by Harry Pennell with assistance from Wilfred Bruce and Victor Campbell, takes a course directly to King Edward VII Land, but like Shackleton three years before, they meet a barrier of dense, impenetrable pack ice. Turning back, they head for Scott's Balloon Bight, arriving there in a gale, late on the evening of February 4, to find that, indeed, it and the nearby bight discovered by Borchgrevink have gone, replaced by a large bay and now the edge of the shelf is considerably farther to the south. It is the Bay of Whales, just as Shackleton had said and Priestley already knew. At least Priestley is glad to have "set the matter at rest finally."

It is just after midnight on February 4, 1911, as the *Terra Nova* edges its way deep into the bay and, yes, there are whales blowing around them. Bruce is on the bridge when he sees the most unexpected of sights. They are at the extreme end of the great Ross Ice Shelf in Antarctica. It is a place that Bruce has already described in a letter to his sister as being the most desolate place in the world. And yet, here, tied up to the ice, is another ship.

As Levick records after he is woken and rushes on deck with the rest, "None of us needed to be told that it was the 'Fram'."

They moor the *Terra Nova* nearby and Campbell, Levick, and Priestley ski over to what they think is the Norwegians' hut, only to realize it is a store. Going over to the *Fram*, Campbell goes aboard and from the lone watchman learns that their hut is actually two miles away from the edge of the ice and that Amundsen is expected at the ship at 6:00 A.M.

———

It is February 4, 1911, 6:30 A.M. Amundsen is driving a dog team and an empty sledge to his ship to pick up supplies. He sees the men driving the two teams ahead of him stop and then wave their arms wildly. When he gets to them, he involuntarily starts gesticulating too: all of them now waving their arms like "incurable lunatics."

> *We had talked of the possibility of meeting the Terra Nova . . .*
> *but it was a great surprise all the same.*

Amundsen heads down to the *Terra Nova* to meet Campbell. He looks older than Campbell expected, a "fine looking man" with "hair nearly white." From Priestley's perspective, however, what impresses him most is the perfect control and ease with which Amundsen works his dog team.

> *I think that no incident was so suggestive of the possibilities*
> *latent in these teams as the arrival of Amundsen at the side of*
> *the Terra Nova. His dogs were running well and he did not*
> *check them until he was right alongside the ship. He then gave*
> *a whistle, and the whole team stopped as one dog.*

The inescapable conclusion is clear to everyone on the *Terra Nova* that morning, not Priestly alone. In what is now, beyond any doubt, an out-and-out race for the Pole, "The principal trump-card of the Norwegians was undoubtedly their splendid dogs."

The exchanges between the men are very cordial. Amundsen invites Lieutenant Pennell, commander of the *Terra Nova*, Campbell, and Levick to come to their hut and base, called Framheim, to have breakfast. It is Levick's turn to be impressed.

> *We found them all men of the of the very best type, and got on very well.*

Amundsen offers that the English can set up their base next to his, and the Norwegian-speaking Campbell is tempted. They can certainly reach King Edward VII Land from here and carry out their ambitious plan for exploring the untouched land. But Bruce and others argue against it on the grounds that, "the feelings between the two expeditions must be strained."

There is nothing more for it. After reciprocating, by hosting Amundsen and some of his men for lunch on board the *Terra Nova*, they cast off at 2:00 P.M., turning their backs on their mission.

And so it is that the three Norwegians have conspired to alter the course that Levick's life takes that day: Borchgrevink had discovered that in this area the barrier (perhaps by virtue of being bent and broken by an underwater island over which it passed) was accessible, then Nansen had given his ship *Fram* to this other Norwegian, this "fine looking man" as Campbell described him, who had sailed it here and set up camp on the only avenue available to the British to carry out their intended exploration.

The *Terra Nova* will now head first to Cape Evans to deliver the news and drop off the two ponies that they have with them, as Scott will need all the help he can get if he is to beat the Norwegians and their dogs. Then, they will head north to Cape Adare.

If Levick is perturbed by the events of February 4, 1911, such a pivotal day in his life, he does not let on. In summing it up in his diary, he writes, "This has been a wonderful day."

PART THREE
CAPE ADARE

Infidelity

The notion that penguins might behave like lotharios, sequentially seducing a bevy of partners, runs counter to the popular image of these famously monogamous and endearing creatures. Moreover, to even suggest that some of them might actually cheat on their partners—have an affair, in our parlance—seems to stretch the bounds of belief, to tar the penguins with our own sins. The view that penguins are more upright than us, morally, if not also in their stance, is as prevalent in science as much as it is in society.

There are several factors that make divorce in Antarctic penguins from one season to the next seem, if not morally admirable, then at least understandable. Their imperative is to breed. Their chances of success are slight. Not only is there high mortality of eggs and chicks, but the great majority of chicks that fledge do not survive to reproduce. In a Darwinian sense, most breeding attempts are dead ends. Storms, predation, the amount of ice cover, the location, and amount of food are all factors that influence breeding success but are variable and largely outside the control of the birds. The one factor that the penguins can control is when they breed. The best predictor of whether a chick will survive to reach reproductive age is how big it is when it fledges. In other words, to have any reasonable chance of being successful, parents need to get cracking with breeding as soon as the conditions are favorable to do so. It would be Darwinian suicide to wait for a previous partner and delay breeding, even if that tardy partner is alive.

On the other hand, once penguins have paired for the season, surely they should stick with it. They don't need to mate together for the rest of their lives, but surely they can and, in their best biological interests, should be faithful for a year? Is that too much to expect?

CHAPTER TEN
THE NORTHERN PARTY

I f the Earth were a body, then Cape Adare in Antarctica would be its genitals, the place where you would be most likely to get screwed. It is windswept, bleak, and dangerous. Hurricane force winds sweep down from the Polar Plateau and sandblast it—rock blast it, really—into submission. Today, it is one of the most out-of-the-way, least visited spots in Antarctica. And for good reason: it is beyond helicopter range of the bases in Antarctica and the sea ice conditions often make it difficult for ships to get into it. Ironically, it is here that Carsten Borchgrevink and his men chose to overwinter on the snowy continent a dozen years before the Northern Party found itself facing the same prospect.

It is February 18, 1911. The *Terra Nova* is once more in Robertson Bay, approaching Cape Adare in the early morning. There is a good deal of pack ice being pushed about by the current and up and down by the swell. The shore itself is encrusted with newly frozen pancake ice, making a landing difficult.

Campbell and Levick really do not wish to establish their winter quarters here. They have read Borchgrevink's disparaging reports of the place, which only confirm what their own eyes can see: they would be trapped on a narrow spit of land, surrounded by mountains and glaciers so ragged and rugged that they would be prevented from exploring the land to the north and west, which is now the target of their new orders. Levick sums their situation up succinctly:

> *On this little patch of peninsular (sic), about a triangular mile in extent, we are absolute prisoners until the sea freezes over in the autumn and allows sledging, as the mountains inland are simply impassable for sledges . . .*

The only possibility to go anywhere from Cape Adare is to winter over and hope that in the spring, the waters of Robertson Bay remain frozen long enough that they will be able to pull their sledges across the ice to their quarry. It is a risky plan, but they are running out of options as fast as the *Terra Nova* is running out of coal. They had spent the previous few days cruising westward from Cape Adare, but were faced with a coastline that consisted of sheer cliffs over one thousand feet high and, as Campbell wrote wryly in his log, "No sign of a possible landing anywhere."

The *Terra Nova* needs to make the long journey back to New Zealand for coal and resupplies and will not return to Antarctica until the following summer. Campbell, Levick, and Priestley discuss their choice: landfall in a desolate and daunting place offering at least a chance of exploring Antarctica westward of Cape Adare or the ignominy of a winter spent twiddling their thumbs in New Zealand. They opt for Cape Adare.

It is difficult getting their stores and the pieces of their prefabricated hut ashore through the swell and the ice, while the *Terra Nova* stays nearby for a couple of days. Borchgrevink's hut is still standing there "in good preservation," as Levick notes, in the midst of the penguin colony that covers the entire spit of land that is Ridley's Beach. Campbell even confesses that he likes the design of the Norwegian hut much better than their own—at least for a small party—although it is already occupied by a molting penguin they christen "Percy." Borchgrevink's hut is dirty and stinks because of the penguins, but the men are forced to stay in it while they erect their own hut next door. Percy, is not to be evicted so easily, however, and he takes up station at the door of Borchgrevink's hut.

Other penguins are attracted by the activity of the human newcomers and naively walk over to investigate. Unfortunately for them, Campbell must make amends after discovering that the sixteen carcasses of mutton they have unloaded from the ship are "covered with green mould."

> *I am sorry to say that a great many visitors we knock on the*
> *head and put in the larder; Percy, however, is sacred.*

Percy stays with them for two weeks, sharing the storeroom of Borchgrevink's hut with the six men who have now changed from the Eastern Party to the Northern Party. One day, suddenly, he is gone. Priestley fears that Percy, having finished molting, has shown up at the hut in his fancy new suit of feathers only to be "taken for a stranger and killed" by the men.

On March 4, their own hut is ready and the men move into their new quarters with much fanfare that includes "a great house warming, gramophone concert, (and) whiskey toddy." Their six beds are arranged in position. On one side—the side with windows—there are the beds of the "officers": Levick, Priestly, and Campbell. On the opposing wall, with no windows and less space, are the beds of the "men": Abbott, Dickason, and Browning. They may very well be in the Antarctic, and even then isolated by even the standards of the Great White Continent, but Campbell is not about to let slip any of the disciplines or structures that in the navy masquerade as procedure but are really just a way of maintaining order.

As the men prepare themselves for the winter among the remnants of the penguin colony, there is nothing at this stage to suggest that

Levick takes any more interest in the penguins beyond their culinary value. Although there is a hint in his reaction to killing Percy's mates, if not Percy himself, that suggests he has a soft spot for the penguins:

. . . dear little things, and I hate having to kill them.

The *Shokalskiy* pulls in to Robertson Bay and I find it exactly as the Northern Party had found it over a century before: a heaving sea covered in loose pack ice surrounds the spit of Ridley Beach making a landing problematic.

Borchgrevink's hut still stands there as solid and sound as it was when he built it. There is a small wooden awning that remains beside it, the leftovers from the Northern Party's hut, which has long been blown to smithereens by the katabatic winds that rip across Ridley Beach from the Polar Plateau. And, in the fate of those two huts, evidenced so starkly on a beach at Cape Adare, surrounded by hundreds of thousands of breeding Adelie penguins, there is everything you need to know about what it takes to be successful in Antarctica and why the race between a Norwegian, brought up living in snow and ice, and an Englishman, with good intentions but no such experience, should turn out as it did.

The seeds of that outcome were already being sown while the Northern Party was sorting out its accommodation at Cape Adare. It was not possible for a party to carry enough food to get to the South Pole and back, and so, during February 1911, Amundsen and Scott's men were out laying depots of food and fuel.

It is February 18, 1911. On the very same day that the Northern Party is coming ashore at Cape Adare, Scott is establishing his farthest south food depot in anticipation of their assault on the South Pole the following spring. It is at 79°28′30″S and they call it One Ton Depot in recognition of the amount of food and supplies it contains. It has taken the depot party of thirteen men, eight ponies, and twenty-six dogs twenty-four painful days to get the supplies this far. The motor sleds had broken down and could not be used for laying the depots at all.

The ponies are what is really holding them back. Gran, the Norwegian skiing expert, whom Nansen had recommended to Scott, is dumbstruck that Scott will not use skis, which he demonstrates can move easily on the snow. But even he is forced to abandon his skis as it proves not possible to lead a pony while on skis. The temperatures drop below -4°F and, with the blizzards they encounter, the wind chill is much colder still. The ponies cannot handle the cold and the men are forced to stop their progress, building snow walls to shelter the ponies as best they can. By contrast, the dogs just settle down under the snow as comfortable as can be. The blizzards are only "a pleasant rest" for the dogs, Scott writes in his diary, "They are curled snugly under the snow and at meal times issue from steaming warm holes." Meanwhile, the ponies are suffering: "so frozen they can hardly eat," observes Gran.

Three ponies had proven so miserable that Scott sent them back early to Cape Evans. And because one of the remaining ponies Scott has with him, Weary Willy, is so profoundly distressed, Scott decides against pressing on further south even though their initial plans had been to site their final depot at 80°S. They are forty miles shy of that mark.

The pony handler, Titus Oates, argues with Scott, saying they should kill Willy and depot the meat for the dogs and men, while making sure the final depot is closer to the Pole. But Scott will not tolerate perpetuating "this cruelty to animals," as he calls it. In a reply that borders on insubordination as much as it does prophecy, Oates says, "I'm afraid you will regret it, Sir."

The journey back is slow and fraught. Along the way, Willy dies.

In fact, of the eight ponies they have started with, only two will make it back to Cape Evans alive: three, including Willy, die because they have become emaciated and physically distressed, unable to cope with the blizzards and the cold.

Birdie Bowers leads a party consisting of himself, Tom Crean, Apsley Cherry-Garrard, and four of the ponies across the sea ice toward Hut Point. They camp on the frozen ice. However, the wind shifts and the ice breaks up into small floes. Bowers wakes and looks out the tent to discover the ice has split apart under the very line where the ponies are tethered. One of them has disappeared, presumably drowned in the freezing waters of McMurdo Sound. At much risk to themselves, the men try to get the ponies and equipment to the solid barrier ice by jumping across floes each time an opportunity arises, slowly making their way to the safety of the barrier. At any time, the wind might shift again, blowing them up McMurdo Sound and to oblivion. Killer whales track their every move. In the end, they manage to get one of the ponies onto the barrier ice, but two fall in the water, and although they retrieve them, the ponies cannot go on. Oates, who has joined them by now, kills the first with his pickaxe, then Bowers must do the same for his pony. It is a grisly business but an extraordinarily lucky escape for the three men who had been trapped on the ice floes with the ponies. Bowers may well lament that he is left "carrying a blood stained pick-axe instead of leading the pony," but the real miracle is that he is left able to consider that at all.

Perhaps the one lesson that all the men, with the possible exception of Scott, seem unable to see past, is that dogs, with proper handling, travel fast and are unaffected by the blizzards, which slow the ponies and lay them low.

To make matters considerably worse for everyone, upon their return from laying the depots, they are met with the news that Amundsen is at the Bay of Whales with his hut well established upon the Ross Ice Shelf and already sixty miles closer to the South Pole than they are. Not only that, because his dogs are inured to the cold more than are their ponies, he will be able to depart ahead of them in the spring. The one lesson it seems that Scott has really learned from the depot laying is that to protect the ponies from the cold, "It makes a late start necessary for next year."

Meanwhile, Amundsen takes seven of his men with him and heads out from Framheim on seven sledges pulled by six dogs apiece. Able to travel at two to three times the speed of Scott and his ponies, the last depot they lay is at 82°S, about 175 miles closer to the pole compared to One Ton Depot.

Amundsen has taken care to mark the route using marker flags and painted food cans placed every mile. He builds large six-foot-high cairns of snow and ice every three miles, in which he leaves a note giving the direction and distance to the next food depot. He establishes more than three food depots for every one the English do. Furthermore, the depots are marked by a line of flags transverse to the route, five miles in either direction. Amundsen is taking out insurance to ensure that even in bad weather, with snow drift and fog, they will be able find both the route and the food depots.

Scott, by contrast, does not mark the route, and his less frequent food depots are marked by only a single flag atop a cairn of snow. By themselves, these might be considered risky as opposed to bad decisions. If there has been a bad decision, it is not laying the final depot, the one farthest south, at 80°S as planned. The decision to place One Ton Depot forty miles short of its intended location is one that will, as Oates predicts, likely come back to haunt Scott later. If there is one thing needed to survive that it is difficult to come by in the Antarctic, it is food.

Penguins are caught in a dilemma whereby they must live in two worlds: using the land for breeding but the sea for feeding. During the summer of '85, one of my PhD students is with me at Cape Bird and we are investigating what the Adelie penguins are eating when they are out in the Ross Sea.

Of course, it is a lot easier to study their behavior on land than it is at sea. It is not like we can travel with them and record what they are eating. Instead, we do it by "reverse feeding": that is, by getting the penguins to throw up what they have already consumed. This is not a technique for the fainthearted or indeed anyone with the slightest sensibilities about the welfare of animals. I daresay Scott would have disapproved.

Essentially, we wait on the shore for the penguins to return to the beach from a feeding trip to sea during the incubation period, and then catch them. That is the easy bit. The "water-offloading technique," as it is called, involves pushing a tube down the penguin's throat so that it passes down the esophagus into its stomach. Sea water is then pumped into the stomach until the bird feels compelled to vomit: at which point the bird is held over a bucket, rather like a drunk puking into a toilet, and the contents of its stomach, which contain what it has eaten over the last twenty-four hours or so, come rushing out. To make sure that the stomach contents are completely flushed out, the process is repeated.

To be honest, I hate sampling the penguins in this way. I am more Scott than Amundsen in that regard. Yet, there is one thing I detest even more: sorting through the vomits. It is not like you end up with a bunch of fresh sea creatures like you might pick up from a fishmonger's. The spew contains the krill, fish, and amphipods eaten by the penguins, all of which are in various stages of digestion. I leave the sorting for my student to do.

The good thing, from a scientific perspective, is that there are hard pieces of the animals that take a long time to digest even after the rest of them have been turned to mush by the bird's stomach juices. These hard pieces can be used to identify the species eaten, the quantity, and even their size and age.

We discover that the penguins are eating mostly a small species of krill, known scientifically as *Euphausia crystallorophias*. This contrasts with the Adelie penguins breeding on the other side of Antarctica, where around the Antarctic Peninsula the penguins are eating almost exclusively a much larger species of krill known as *Euphausia superba*. Krill are small crustaceans that look like mini lobsters. They form part of the zooplankton, and in the summer in the Southern Ocean, they occur in enormous numbers. In fact, *Euphausia superba*, despite their small size, are estimated to collectively comprise nearly four hundred million tons of krill in the Antarctic's ocean, making it perhaps the largest biomass of any species on the planet. Despite their collective mass, the penguins must find, catch, and eat a lot of individual krill if they are to derive enough energy to sustain their breeding. This is especially so for our penguins in the Ross Sea, which are reliant mainly on the much smaller species of krill. One of the birds we sample contains the remains of 41,938 individual krill in its stomach.

While the water-offloading technique looks somewhat like water-boarding and akin to torture, the truth is that it does not harm the penguins in any way, even though it is no doubt unpleasant for the birds. Penguins are biological machines that are designed, essentially, to withstand long periods of fasting as they must nest on land away from their food sources. Losing what they have eaten during the last few hours will have little impact on their survival. Even so, I, for one, am glad that in the world of penguin research this technique has now been largely superseded by other techniques that can deduce much about what penguins have eaten from analyzing their feathers or their poo. Though, I will leave sampling the latter to my students too: bodily excretions, be they vomit or urea, are just not my cup of tea.

Knowing *what* penguins eat seems less important to understanding the factors that influence their breeding success than knowing *where* they eat. My data, collected during that first summer in 1977, had shown that if feeding birds spend too long away at sea feeding, their partner on the nest will desert their eggs, or their chicks will hatch and starve to death because the partner on the nest cannot feed them. If the penguins are feeding close to the colony during these long incubation-period feeding trips, their tardiness would be inexcusable. Hence, it seems likely that they must be traveling long distances to get their food, and as a consequence, the difficulties of coordinating nest relief for a penguin couple will become magnified.

Distance has a way of magnifying problems. It doesn't just make the heart grow fonder, it can make the mind grow angrier.

It is March 28, 1911, and Douglas Mawson is on the other side of the world, in London, making preparations for his expedition to Cape Adare when the *Terra Nova* reaches New Zealand's Stewart Island. This sparsely populated island at the bottom of New Zealand is covered with luxuriant rainforest and is a complete contrast to the white treeless continent from where the *Terra Nova* has come. The news it relays to the rest of the world is, however, anything but colorless: Amundsen is in the Bay of Whales; Scott is in McMurdo Sound; next summer there will be an all-out race between the two parties to get to the South Pole.

Yet the thing that perturbs Mawson most in the news is that the Eastern Party has now given up their proposed exploration of King Edward VII Land and have, instead, been landed at Cape Adare. Mawson cannot hide his anger when approached by a reporter from the *Daily Mail*:

> *It surprises me very much to hear that Captain Scott has landed a party at Cape Adare in face of the agreement between us and in view of the information I gave him . . . Captain Scott personally wrote to me the last thing before leaving Australia and asked me to furnish him with full details of my plans, and this I willingly did, giving him all particulars, including the statement that I intended to land at Cape Adare.*

There is so much irony in this. Scott had been distraught to learn that Amundsen had based himself in the Bay of Whales for his own attempt to get to the South Pole. Cherry-Garrard was with Scott when he received Campbell's news about the Norwegians:

> *For an hour or so we were furiously angry, and were possessed with an insane sense that we must go straight to the Bay of Whales and have it out with Amundsen and his men . . . we felt, however unreasonably, that we had earned the first right of way.*

Yet Scott thinks it perfectly okay to have his men usurp the territory that Mawson had claimed for his own expedition. Mawson has kept Scott informed of his plans, and now Scott has done a "Shackleton" on him.

Kathleen Scott, upon seeing Mawson's anguish laid out so bare in the newspaper, writes to Mawson to soothe his hurt and, later, invites him to lunch at her home. It does not assuage the anger Mawson feels toward Scott's actions, but it does make him feel better about Kathleen. And, it has to be said, she feels a warmth for Mawson too. At that moment, her husband is in the Antarctic entering a dark cold winter, while Paquita, by now his fiancée, waits for Mawson in Australia.

Had I been in my generator box, observing Kathleen Scott and Douglas Mawson during their lunch, I might have described them, in penguin terms, as approaching each other cautiously and bowing. That is, in the very preliminary stage of courtship when a female investigates a male performing his ecstatic display. Most often it goes nowhere: she turns and runs away. Most often.

CHAPTER ELEVEN

THE WORST JOURNEY

Of course, not everything we think we know about penguins turns out to be true.

It is June 22, 1911, and men sitting in three huts dotted around the Ross Sea celebrate Midwinter's Day. Amundsen and his men are at the Bay of Whales, comfortable in their warm hut with its adjoining network of under-ice rooms carved out of the barrier ice. There is ample food, a library of three thousand books, a piano, a gramophone, and even a sauna. Even given the much more modest comforts of the Cape Adare hut and their normally regimented routine, according to Levick, they manage to celebrate in style with champagne, brandy, cigars, and "an extended sing-song."

Apsley Cherry-Garrard is at Cape Evans. The young, nearsighted adventurer was employed by Scott as assistant zoologist—despite having no training in that regard—only after he offered to forgo a salary and, instead, pay Scott £1,000 toward the expedition's costs. He describes the scene that night:

> *Inside the hut are orgies. We are very merry—and indeed why not? The sun turns to come back to us tonight, and such a day comes only once a year.*

I went to Scott's hut at Cape Evans for the first time in 1985. The last word I would use to describe its long dark interior is "merry."

Certainly, if anyone has cause for not being merry that Midwinter's night, it is Cherry-Garrard and two of his companions: the doctor Edward Wilson and Birdie Bowers. They are preparing to leave five days later to march to the Cape Crozier Emperor penguin colony to collect eggs, the embryos of which, Wilson surmises, could prove to be the missing link in explaining the evolution of birds from reptiles.

Wilson had already determined during the Discovery Expedition that Emperor penguins, unlike the summer-breeding Adelie penguins, bred during the Antarctic winter. At the turn of the 20th century, these large flightless birds are thought to be the primitive precursors of flying birds. Wilson reasons that if they could just collect some eggs from the winter-breeding Emperors, the eggs should provide insights into the evolution of all birds. There is a generally held belief among scientists at this time that, "embryology recapitulates phylogeny." Or rather, in something approaching English: when an animal is developing in the womb or the egg, it goes through developmental stages that correspond with its evolutionary past, becoming ever more complicated as it matures.

In 1911, the only known place where Emperor penguins breed is on the sea ice off Cape Crozier, on the other side of Ross Island from the Cape Evans hut.

It is pitch black when the three men leave, hauling two sleds weighing 759 pounds, in temperatures that rarely go above -50°F. Their journey to Cape Crozier in winter will be immortalized by Cherry-Garrard in *The Worst Journey in the World*, the book that will establish a longing in me to go to Antarctica, although, in hindsight, its descriptions are so shocking in their brutality that it seems unfathomable to me now that they did not frighten me away.

> *It took two men to get one man into his harness, and was all*
> *they could do, for the canvas was frozen and our clothes were*
> *frozen until sometimes not even two men could bend them*

into the required shape . . . Once outside, I raised my head to look round and found I could not move it back. My clothing had frozen hard as I stood—perhaps fifteen seconds. For four hours I had to pull with my head stuck up, and from that time we all took care to bend down into a pulling position before being frozen in.

It takes them nineteen days to cover the sixty-seven miles of crevasse-filled ice to Cape Crozier.

The horror of the nineteen days it takes us to travel from Cape Evans to Cape Crozier would have to be re-experienced to be appreciated; and any one would be a fool who went again: it is not possible to describe it. The weeks that followed them were comparative bliss, not because later our conditions were better—they were far worse—because we were callous. I for one had come to that point of suffering at which I did not really care if only I could die without much pain.

It is not just the terrain with its crevasses or crystallized snow that makes it like pulling their two sledges through sand—meaning that they must take one sledge forward, then go back and get the other, walking three times the distance that they manage to bring themselves closer to the penguins each day, which often means making only a mile or two of progress. It is not just the darkness, which means that they cannot see the pitfalls ahead of them, nor the way back to the other sledge. It is the bitter cold, when all their clothing and sleeping bags are drenched and stiff with frozen sweat, frozen breaths, and frozen snow. "They talk of chattering teeth," writes Cherry-Garrard, "but when your body chatters you may call yourself cold." The temperatures get as low as -77.5°F.

Once at Cape Crozier, they build an igloo of rocks and snow on an exposed moraine using a piece of canvas as its roof. They are some eight hundred feet above the sea ice on the slopes of Mount Terror.

It is 1985. I am at Cape Bird studying the penguins when a helicopter arrives to pick up two surveyors and transport them to Cape Crozier. The pilot suggests that there is space in the "helo" for me and one of my assistants, and "why don't you come along for the jolly?" as such perks are called on the ice. He says he can bring us back later, as he needs to come back to Cape Bird. We do not have permission from the New Zealand operators at Scott Base for such a mission, but this is something being offered by the Americans. "Never wait for the second shuttle," is something another helicopter pilot has said to me and it resounds in my head. We climb aboard.

At Crozier, we land near the knoll where Cherry-Garrard, Wilson, and Bowers built their igloo of rocks and snow nearby. All that remains is an oblong rectangle of stacked black boulders partially covered in snow, knee-high. It looks like it gives way less shelter than even my generator box. The slope itself stretches away to the ice-covered sea. The patchwork of snow and black volcanic rock leaves no doubt that even if Mount Terror is no longer active, it had certainly been so once. To the right is the great white expanse of the Ross Ice Shelf and, where this giant glacier comes into contact with the sea ice, where the surface of the sea has frozen during winter, the sea ice has been buckled by the pressure coming from the slowly moving shelf and been forced up into a jumble of pressure ridges like miniature mountains. Beyond them, perhaps four miles away, there is a dark patch on the sea ice, which, when I use my binoculars, I can just make out is a congregation of Emperor penguins and their large gray fluffy chicks. The ice is stained from their guano and, I suppose, their dirty guano-covered feet.

Even then, even in the full daylight and comparative warmth of the summer, it would not have been easy to reach the penguins. As we stood there contemplating that, suddenly the wind changes unexpectedly and jumps in velocity to over twenty knots, gusting much higher. The pilot has shut down the helo and has to act quickly to get the rotor blades going and lift off, otherwise we will be forced to take shelter near the knoll, just as Cherry-Garrard and the others had, and our pilot clearly is in no mood to engage in such discomfort.

We lift off with a sudden jerk eastward, propelled by the wind as much as the engines. There is no way that we can beat back into it to return to Cape Bird, so the pilot heads southward, taking what little shelter he can in the lee of Mount Terror, finally dropping my assistant and me at Scott Base, much to the consternation of the commanding officer at the base—I had broken the first rule of Antarctic travel: get permission first.

The storm rages for three days before the wind finally dies down enough to enable us to go back to Cape Bird, tails between our legs.

⚬

Wilson, Bowers, and Cherry-Garrard make their first attempt to get down to where they believe the Emperor penguins are breeding on July 19, 1911, but they become entrapped within the mess of crevasses and pressure ridges, which in the almost total darkness are impossible to negotiate. They can hear the penguins, but are forced to return to their camp. The next day they try again, this time near the middle of the day when the darkness is not so "pitchy black."

> *After indescribable effort and hardship we were witnessing a marvel of the natural world, and we were the first and only men who had ever done so . . .*

The colony of penguins seems to be in some strife. There are only one hundred birds there compared to two thousand or so when Wilson had been there on the Discovery Expedition, and, of those, only about a quarter or less are incubating an egg. Many eggless birds, so desperate to incubate, so primed by their hormones and instincts, are incubating egg-shaped blocks of ice instead.

The men collect five eggs, kill and skin three penguins, and then make their way gingerly back to the igloo in the cold and the dark.

*. . . we on this journey were already beginning to think of death
as a friend. As we groped our way back that night, sleepless, icy
and dog-tired in the dark and the wind and the drift, a crevasse
seemed almost a friendly gift.*

The eggs are in the insides of their fur-lined over-mittens, which they
carry tied about their necks but by the time they get back to the igloo,
the two that Cherry-Garrard carries in his mittens have been broken.
They pickle the embryos from the remaining three eggs in alcohol.

The next day, the weather takes a turn for the considerable worse.
A storm of hurricane proportions rips the canvas roof away from their
shelter, the sides cave in, and some of their belongings, including their
only tent, is blown away.

*. . . it was blowing as though the world was having a fit of
hysterics. The earth was torn in pieces: the indescribable fury
and roar of it all cannot be imagined.*

They lie there for two days in their sleeping bags, covered in snow
drift, their voices unable to be heard above the roar of the wind, as certain of death as it is possible to be. Then, after such a long time without
hope, there is a sudden lull in the storm; a lull long enough for Bowers
to find and retrieve their tent. Somehow—even Cherry-Garrard in his
book seems unsure how—the three men and the three eggs make it to
the safety of the hut at Hut Point, the one that Scott had built during
his first expedition. The worst of the world's worst journey has come
to an end.

———

I am following the surprisingly quick footsteps of the waistcoat-wearing
Douglas Russell as he leads me into the bowels of the Natural History
Museum's storage facilities at Tring. It is a veritable mall of birds' eggs
and nests, as well as being a mortuary for the carcasses of animals

collected during a time when it was thought that the most essential item needed to study biology was a gun. It would take hundreds of years before recognition of the irony of inflicting death to learn about life should bring a halt to such practices.

Douglas opens a large gray metal cabinet and hauls out a thick-walled jar the size of a jug of beer. At its bottom, pickled in the yellowish alcohol, lies one of the hard-won embryos from the "winter eggs." It is pale and remarkably big; about the size of Douglas's thumb. Most of the head seems taken up by two enormous black eyes. If the embryology of Emperor penguins can tell us anything, I reckon, it is that this nascent creature had been on its way to becoming a bird with damn good eyesight.

Douglas tells me that of the million-plus exhibits held at Tring, more people come to see the "winter eggs" than all the other specimens combined. He fetches a plain, oblong cardboard box and ceremoniously removes its lid revealing the two eggs it contains. The very eggs that Cherry-Garrard had delivered personally to the museum and for which he, Wilson, and Bowers had suffered so hard and for so long. The third egg, Douglas explains, is currently in an exhibit at the Natural History Museum in South Kensington.

I am stunned; more by what the eggs represent rather than how they are presented. They are large, oval, whitish, and with unexpectedly large and irregular holes through which Wilson had pried the embryos. They seem so unprepossessing, so lacking in any quality that could make men endure such extreme hardships and flirt so closely with death.

Douglas asks me if I wish to photograph them. It is only when I am perched over them that he enquires as to whether I am sure that my lens is attached securely to my camera. I experience a moment of pure panic. My camera is directly above the eggs, a foot away from them, no more. I don't even have a wrist strap attaching it to me: one cough or a slip, and I will destroy the most famous, most visited items in the collections of the Natural History Museum. I click the shutter quickly, a couple of times, then jerk my camera to the side.

For all their sentimental value, the eggs proved to have virtually no scientific value. Soon after they were deposited in the museum, science realized that penguins were not the original primordial birds but, instead, were derived from flying birds and, even then, relatively recently. In fact, after some cursory initial attention, it was decades before anyone at the museum even bothered to look at the eggs in any serious way.

Yet, if Cherry-Garrard and the others had risked so much for what science treated as being of so little value, I find some enlightenment in his first-ever descriptions of Emperor penguins breeding, as he puts it, "in the middle of the Antarctic winter with the temperatures anywhere below seventy degrees of frost, and the blizzards blowing, always blowing." He notes that the birds do not have a nest site but "shuffle along" the ice, carrying the egg on their feet.

Where he gets things wrong is that he assumes that these birds carrying the eggs are the females. Nothing exemplifies why Cherry-Garrard is not destined to be the world's first penguin biologist more than this observation of the incubating penguins:

> In these poor birds the maternal side seems to have necessarily swamped the other functions of life. Such is the struggle for existence that they can only live by a glut of maternity, and it would be interesting to know whether such a life leads to happiness or satisfaction.

I cannot help but think of the irony of the comparison with Levick. Cherry-Garrard was employed as the Terra Nova Expedition's assistant zoologist but he eschewed the style of the dispassionate scientist for that of the breathtaking writer, a man capable of wringing emotion from words in a way that would inspire a young boy in New Zealand half a century later. Whereas Levick, the surgeon who had swapped his double-barreled shotgun for Wilfred Bruce's fountain pen as he wished to become a writer, should prove to be so objective and methodical an observer of nature that he would explore the scientific ground later covered by that same boy.

While admittedly no scientist, Cherry-Garrard's wistful reflections are understandable given his own struggle for existence, and not just that of the penguins, during his journey to see them:

> I might have speculated on my chances of going to Heaven; but candidly I did not care. I could not have wept if I had tried . . . Men do not fear death, they fear the pain of dying.

It is one of the most gobsmackingly remarkable features of this bird, which breeds in the dead black of the Antarctic winter, that the male alone should incubate the egg for two long months with "the blizzards blowing, always blowing" against his devoted back until the chick hatches. Setting aside that they were the fathers and not the mothers that Cherry-Garrard thought he was observing, his observation that they shuffled about not attached to any particular nest site explains a lot about the mating of Emperor penguins that it would take the better part of a century for scientists to understand.

Wilson may have been wrong about the scientific importance of the embryos that were the motivation for the winter journey to Cape Crozier, but Cherry-Garrard was not wrong about his observation that they had witnessed "a marvel of the natural world." The Antarctic is such a hostile place in winter that it seems incredible that any animal should choose to live there at that time, let alone breed. And it is true that all the birds and mammals that make the Antarctic continent their home in summer—when there is twenty-four-hour daylight and the seas are brimming with krill and the greatest abundance of life found anywhere on the planet—get the hell out of there at the first signs of winter. All, that is, except Emperor penguins.

It is 2005. I go to the theater to see a film about Emperor penguins. The story of their extreme lifestyle has gained international attention with the release of a feature-length documentary, *March of the Penguins*. It is immensely popular and becomes the highest-grossing documentary ever made, pulling in over $127 million at the box office.

The basic story of the Emperor penguins as told in the film is entirely accurate. They breed in the heart of the Antarctic winter. The pairs form in April. A single egg is laid and carried on the feet of the father for some two months while the mother goes to find food at sea. It is completely dark. Temperatures can be as low as -60°F, and winds of over one hundred mph can take the wind chill factor down to levels that make the penguins' persistence seem more like lunacy than some marvel of the natural world. The mother returns just as the chick hatches, and then, together, mother and father alternate turns, with one feeding and brooding the chick while the other goes to sea to get food.

This cinematic story of a struggle for survival against the odds may have resulted in little more than fascination with one of the peculiarities of biology were it not for the spin the film puts on it: the producers sell it as a love story. As the posters proclaim, "in the harshest place on Earth love finds a way." The film makes love the enduring, most highly valued commodity of penguins.

The Christian Right in the United States embraces the story. Conservative Christian groups co-opt Emperor penguins as evidence for "Intelligent Design" and family values. As one of their publications puts it:

> *. . . the film reinforces monogamous heterosexual nuclear family structures as an innate and desirable part of life.*

Nothing can be further from the truth. Emperor penguins are by far the least faithful of any penguins, with 85 percent divorcing their partner and taking up with a brand-new one from one year to the next. Without a nest site to act as a rendezvous point from one year to the next, as we observed to be so important for the reunification of Adelie penguin pairs, and with pressure to initiate breeding as soon as possible, it is more luck than love that enables the relatively few pairs of Emperor penguins to reunite from one year to the next. Rather than being icons for love, Emperor penguins should be the patron saints of divorce.

And yet, the impression persists today—as much as it did during the winter of 1911, when Levick sat on his bed by the darkened window at Cape Adare, dreaming of writing a novel—that penguins mate for life.

CHAPTER TWELVE

THE RELUCTANT
PENGUIN BIOLOGIST

Murray Levick is unaware of the calamitous events taking place at Cape Crozier beyond his darkened window. He and the rest of the Northern Party spend the winter at Cape Adare making preparations for their own sledging journeys, which will begin in the spring. That is, until calamity strikes their party again in its oddly familiar way.

Until now, Levick has done little by way of his secondary function as a zoologist. Instead, he has concentrated mainly on his photography, taking over Borchgrevink's hut for a darkroom. What notes he writes in his big blue Zoological notebook are perfunctory at best.

When they had arrived at Cape Adare, it had been near the end of the penguins' breeding season. Most chicks had fledged, and the adults had either moved elsewhere to molt or hung about at Cape Adare waiting for their new set of feathers to grow. Levick noted that there were only about 1,500 Adelie penguins remaining in the colony on February 18. The remaining chicks, which had yet to fledge, seemed unlikely to survive: they begged from any adult and seemed, to Levick, to have been abandoned by their parents. By March 12 there were no fledgling chicks left, leaving only about three hundred adults in the colony, including just thirty that had still to complete their molt. The last penguins he saw were on April 6: six of them. He killed four of them for food. Not exactly the portents for becoming the world's first serious penguin biologist.

He does not see another live penguin until September 19 when four Emperor penguins visit Cape Adare. As he writes, in a way that appears starker when written in his cursive script in the blue-black ink, "I killed them all."

Whatever reticence Levick may have admitted to killing penguins seems well and truly behind him. Yet there are still the hints of a penguin scientist with a moral heart that whiles away the dark winter hours at Cape Adare (or "Cape Adair" as he, curiously, spells it).

After his initial observations of the dwindling penguin numbers, he wrote in his Zoological Notes mainly about three species of petrels, and some protozoans that he recovered by melting ice, complete with colored pencil drawings. He shows no interest in penguins until August 12, when he dissects a female fledgling penguin that he had found frozen two days earlier. It is completely emaciated, containing no fat and its stomach is full of "surprisingly large" basalt stones and nothing else. The lot of a late fledgling chick is evidently not a kind one.

But Cape Adare has its own peculiarly unkind fate reserved for the six men that occupy the new hut on Ridley Beach that winter.

It is the evening of August 15, 1911. A gale that has been blasting Ridley Beach increases in intensity to hurricane proportions. In an environment where its human inhabitants have rarely known a day without a fierce wind blowing, Dickason describes it as, "the hardest blow we have had."

It is freezing inside the hut, even with the fire going, and the men get into their eiderdowns. Outside the hut is being pelted with wind-driven rocks and Dickason must crawl on his hands and knees when taking the meteorological measurements. The men fear, as Cherry-Garrard and his companions had feared a month earlier when lying in their igloo during a similarly ferocious storm, that the roof may not hold. Indeed, the roof is blown off their storeroom, which had been Borchgrevink's storeroom originally. Their meteorological station and many of its instruments are destroyed. However, all that pales into insignificance for them when compared to the most devastating consequence of the storm.

In the morning, Dickason goes outside the hut only to return quickly with the news that the sea ice, which had formed in Robertson Bay over the winter, has been completely blown away. It is Priestley who sums up the gut-wrenching implications of this for them:

> . . . *we gazed seaward this morning and realized the astounding fact that the sea ice beyond the bay, our only hope for any future sledging, had gone out during the night.*

Once more their mission to go exploring has been thwarted. First at the Bay of Whales, that area of the Ross Ice Shelf discovered by the Norwegian Borchgrevink and subsequently occupied by his childhood playmate. Second, here at the place where men first set foot on Antarctica and overwintered on the continent: and yes, that Borchgrevink again. The guy whose storeroom they have usurped seems to haunt their every move. The sea ice had been their only plan, their only possible route to the northwest, their only possible pathway to map and explore hitherto untouched areas of the Antarctic continent—areas that are now destined to remain unexplored, untouched, and unmapped, while they remain marooned on a tiny triangular spit of land about two square miles in area until the *Terra Nova* can come back to collect them sometime in the New Year.

Cape Adare sits at the tip of a long needle-like point of land, some twenty-five miles in length, that forms the eastern side of Robertson Bay. Directly opposite Ridley Beach, some twenty miles away on the western side of Robertson Bay, is Cape Wood. That is where the Northern Party had hoped to begin their exploration of the Antarctic coastline to the west of there. Crossing the ice-covered Robertson Bay had been their route to

Cape Wood. Yet, even as Levick acknowledges that the storm has delivered "the most awful blow to our hopes of sledging along the coast," had it happened ten days later when a sledging party consisting of Campbell, Priestly, Abbott, and Dickason was scheduled to be out on the ice attempting to get to Cape Wood, they "would have certainly been dead men."

If there is a sliver of hope left for the Northern Party and their quest for exploration, it is that a narrow strip of sea ice still hugs the coastline to their south. Rather than going directly to Cape Wood across Robertson Bay, it may be possible to make their way, cautiously, around the coastline, pulling sleds along the sea ice that remains stuck to the shoreline. It will make for a much longer journey and one where they will need to be ever mindful that the remaining ice does not break out too.

It is September 8, 1911. While Levick and the other members of the Northern Party have been preparing to head out around the edges of the recently ice-free Robertson Bay, at Framheim, Roald Amundsen has been desperate to get his push to the South Pole underway. He is worried that Scott may start early, and while the ponies are probably unable to get underway when the air is still frozen with the brutal winter cold, the motor sledges might give Scott an advantage in that regard. Originally, Amundsen had planned to leave on November 1, but his worry about his rival causes him to move the date forward, to the day when the sun will make its first appearance after a winter of darkness: August 24. However, Hjalmar Johansen, the veteran who had traveled with Nansen in a bid to get to the North Pole and had been beaten back after departing too early when conditions were too cold, disagrees.

> We cannot leave as long as the temperature keeps so low . . . it will be terrible for the dogs.

During the days leading up to their planned departure, the temperature stays below -58°F. Even so, Amundsen orders that the sledges, which have been packed and waiting for a month already, be moved outside in preparation for their departure. The cold and wind does delay them, but on this day, September 8, 1911, with the temperature now up to -35°F, they set off: seven sledges, eight men, and eighty-seven dogs. One dog needs to be shot soon afterward because she comes into heat, causing chaos among the other dogs. For the first few days, the going is good and they cover a satisfying fifteen miles each day. But then, the temperature drops from uncomfortably cold to life-threateningly cold, even with all their "eskimo" gear. It is -69°F. Amundsen is forced to accept that they have begun too early. The dogs are suffering with frostbitten paws. Two dogs freeze to death when they lie down. The men are not much better: several of them have frostbitten feet.

On the return to Framheim, with more storms and cold weather threatening, Amundsen orders that they cover the last forty miles in one go. They had previously unloaded much of their food and gear at their 80°S depot, so the sledges are light and the dogs are keen to get back now they sense that they are heading for home. Amundsen and two others set off first with two of the sledges, covering the forty miles to Framheim in just nine hours. Those behind him do not fare nearly so well: their feet are frostbitten and their dogs are faltering. Two arrive a couple of hours after Amundsen and then another sometime after that. However, the last two, Johansen and Kristian Prestrud, are still out on the Ross Ice Shelf—the barrier ice—without food or fuel.

Prestrud's dogs are unable to pull. He is on skis now, with badly frostbitten feet, moving awkwardly and close to exhaustion. Johansen stays with him and together they press on, arriving at Framheim eight and a half hours after Amundsen, in darkness and in fog, with the temperature now at -60°F. They have had nothing to eat for over nineteen hours and are well and truly on the other side of the line that normally demarks death.

At breakfast, Johansen berates Amundsen for going off and leaving his men.

I don't call it an expedition. It's panic.

Faced with such a challenge to his leadership, with his aim of becoming the first man to get to the South Pole hanging in the balance, Amundsen acts decisively. He tells Johansen and Prestrud that they will now not be part of the Polar Party. They must, instead, explore King Edward VII Land, the place that had been the initial object of the orders given by Scott to Campbell, Levick, and the rest of the Eastern Party. Adding salt to the wound, the young and largely inexperienced Prestrud will be in charge.

Johansen refuses, demanding a written order, which Amundsen duly gives him that evening:

> *I find it most correct with the good of the expedition in view—to dismiss you from the journey to the South Pole . . .*

Johansen is defeated. He becomes morose and a gloom settles over Framheim. Amundsen will barely speak to Johansen again. Amundsen is desperately worried about Scott's motor sledges, but he dare not depart again until his men's frostbitten feet are better, four of whom are bedridden for ten days. He asks one of them to accompany Johansen and Prestrud to King Edward VII Land.

It is September 8, 1911. On the same day that the Norwegians are setting out on their thwarted attempt to get to the South Pole, the six men of the erstwhile Eastern Party shut up their hut at Cape Adare and, instead of King Edward VII Land, which, ironically, the Norwegians will now have for themselves, they set off with much more modest targets in their sights. Levick and Browning pull one sledge. They are to go only as far as Warning Glacier—due south of Cape Adare—to do some photography, while the other party, pulling two sledges, carries on to Cape Wood and then explores the coastline beyond that to Cape North. Dickason, unnerved by the storm that took away most of the sea-ice, leaves a note to the effect that should another storm take away the ice they are traveling on, "Rings" is to pass on his diary to his mother.

Levick and Browning set up their tent on the sea ice in front of the glacier, but that night they get hit by just such a severe storm and by the second night, they realize that the ice on which they are camped is cracking. Levick describes it as "the most trying night I have ever spent" with the "ice rising and falling under us." It is too windy to attempt to shift their tent to the shore, so Levick and Browning take turns keeping watch, ready to abandon the tent and carry their sleeping bags, some food, and a Primus stove if they must make a run for it. In a lull, they manage to get out, grab some quick photos, and then retreat back to their hut.

Meanwhile, Campbell and the others are not affected by the storm even though they are only a few miles away. However, as they approach Cape Wood, the going gets really hard: they sink to their waists in heavy snow drift. While they manage to get to Cape Wood, they realize that they do not have enough food left to do any exploration. They depot what food they have left and return to the hut, arriving there on September 18.

Campbell's plan is to set out again on October 4, 1911, this time carrying more food. Levick and Browning accompany them part of the way once more. Thereafter, Levick takes photographs of the glaciers and Admiralty Range of mountains that can be seen from around the edges of Robertson Bay to supplement the work Priestley is doing on geology. They have instructions to kill a seal and leave the meat in a cave for Campbell's party to use on their return leg. On October 9, they see a female Weddell seal that has given birth to a pup near Abbey Cave. The next morning, they kill and butcher both. "A heartless business," as Levick calls it, but he justifies it as a kind of necessary cruelty.

By contrast, he is not prepared for the evidence of unnecessary cruelty he finds later that same day. He and Browning have managed to pull their sledge as far as Duke of York Island, at the head of Robertson Bay. They set up camp on the sea ice but in a valley coming down to the ice, there are signs of an Adelie penguin colony. Levick sets off by himself, wearing just his finnesko—a soft boot made of reindeer skin—on the stony soil. The going is difficult and he does not wish to travel too far. Even so, he is brought to an abrupt halt. As he describes it in his Zoological Notes:

A dead black throated penguin lay with a rope tied round one of its legs, the other end being made fast to a large stone. The poor bird had evidently died of starvation after many days of struggle to get away back to the sea, for it lay in that direction, on its breast with its head stretched seawards, at the fullest extension of the rope. Its breast bone stuck out in its emaciated breast.

This loathsome little atrocity can only be attributed to one of the members of the Borchgravinks (sic) party who visited Duke of York Island ten years ago, and is worth recording if only as scientific evidence of the wanton cruelty to which some men can descend.

Onboard the *Shokalskiy*, I have with me a copy of Borchgrevink's book *First on the Antarctic Continent*. Indeed, Borchgrevink had made a sledging trip to Duke of York Island and camped in Crescent Bay, just as did Levick. Crucially, this was in December 1899 when the penguins would have been ashore breeding. Borchgrevink makes no mention of the incident of tying a penguin up, or even, the presence of penguins at Duke of York Island.

However, he does write about an event that took place soon afterward on December 21:

. . . the Finn Savio conceived rather a good idea of amusing himself; he caught a penguin, made a string fast to its legs, took it on board his kayak, and used it as motor power. The penguin dived and pulled Per and the kayak about at a great pace . . .

Borchgrevink never traveled anywhere without his two Finns. Hence, it seems likely that it was Per Savio who had tied up the penguin at Duke of York Island as a sort of precursor to his later means of "amusing himself."

⸻

How deeply this atrocity affects Levick is hard to judge. Perplexingly, he seems not to have mentioned it to the other men and, like other material that is written in his Zoological Notes, despite his stated intention to record the event for scientific posterity, he keeps it hidden.

It is on the day that Levick and Browning get back to the hut, October 13, 1911, that he sees the first Adelie penguin arriving at Cape Adare for the new breeding season. Levick records this in his big blue-covered book of Zoological Notes, underlining it with a flourish of his fountain pen as if to emphasize its importance. Indeed, by the time Campbell's party gets back to the hut one week later, as Priestley notes:

> *Levick had started on a series of systematic notes, which are probably the most thorough that have ever been made on the habits of the Adélie penguin.*

Levick's path forward is now clear: he should study the penguins.

> *Have been reading up all I can find about penguins . . . My great ambition now is to work them up thoroughly and write a book on them when I get back.*

Stranded at "Cape Adair," with time on his hands, Levick has resigned himself to conducting research on the penguins as a way to fill in the days. He photographs them and makes copious notes in what will be, as Pricstley has already realized, the first systematic study of the breeding behavior of any penguins.

The men are left in no doubt as to Levick's rigor as a scientist. He has provided a notebook in the hut for the men to record any interesting zoological observations about penguins, other birds, seals, whales, or anything else. He prefaces the notebook with the following rules:

1. *Never write down anything as a fact unless you are absolutely certain. If you are not quite sure, say 'I think I saw' instead of 'I saw,' or 'I think it was' instead of 'It was,' but make it clear whether you are a little doubtful or very doubtful.*

2. *In observing animals disturb them as little as possible. This especially applies to the arrival of the penguins, as it is most important to allow them to settle down naturally without interference from us, and to the giant petrels, which became wilder last autumn after we had hunted them.*

3. *Notes on the most trivial incidents are often of great value, but only when written with a scrupulous regard to accuracy.*

N.B.—Please remember that we have every reason to believe that birds feel pain as much as we do, and that it is well worth half an hour's laborious chase to kill a wounded skua rather than to let it die a slow death.

The instructions say more about Levick than any letters after his name might have. Untrained as a scientist, he nevertheless possesses the mind and methodology of a scientist. It is one that he had already displayed when studying brucellosis in the Mediterranean, and, it is married to a heart filled with respect for his study animals and their welfare.

Yes, he kills seals and penguins for their meat; but that is a regrettable necessity in his mind. What he cannot abide is inflicting unnecessary suffering on animals. And certainly not for fun. In that regard, he is the polar opposite of Per Savio and Carsten Borchgrevink. Causing animals to suffer does not bring him amusement and he thinks much less of the men who would perpetrate what he calls the "scene of tragedy I saw."

In addition to his objectivity, Levick's Zoological Notes reveal that he sets up the sort of experimental observation that is not just years ahead of its time, but decades.

He selects two small subcolonies (using the terms common in his day, he refers to them as "colonies"), "each with about fifty couples, for special observation." He marks the subcolonies with red flags on bamboos "so that they shall be in no way interfered with by any of our party collecting eggs etc." The first subcolony, he calls Group A. It lies on a small rounded knoll surrounded on three sides by a shallow lake that is beginning to thaw. "Being isolated from the main crowd," he suggests, "it will be easy to watch them specially, and get to know habits and characteristics of the various couples individually." What's more, he foreshadows my study seventy-three years later by marking the breasts of five pairs in Group A with bright red paint. One of the supposed pairs turns out not to be a pair: instead, each has its own partner. This makes six nests in all where at least one of the pair is marked individually. Levick is not finished yet. He marks each of their nests with large stones on which he writes the number of the nest: one through six. In addition, there is an easily identifiable injured bird in a nest on the periphery of Group A, and so he includes that nest also as part of his intended, detailed systematic observations.

The experimental protocol at the second subcolony, Group B, is not so well thought through. Rather than decades ahead of its time, it reflects more the Victorian attitude to studying nature: that nothing beats collecting and killing. This subcolony is situated near the meteorological screen and, again, Levick marks the breasts of several pairs in it. However, this time his "intention is to remove all the eggs from this group as they are layed (sic)." He takes four away that are present on the first day and he seems well pleased that, "A couple of minutes after I removed the eggs, the owners seemed to have forgotten the incident entirely." He may well have been trying to determine whether the birds would re-lay (a common feature of some birds that lose an egg soon after it is laid) but, really, such a manipulation on such a large scale was always going to cause massive disturbance. Perhaps for this reason, Group B does not feature much at all in his Zoological Notes.

Of course, when Murray Levick finds himself, the reluctant penguin biologist, standing on Ridley Beach facing a summer studying

the penguins, he cannot have known anything about the sexual misdemeanors that are about to be unleashed by the miscreants at his very
feet. At that point in time, the behavior of penguins has never been
studied in any detail. However, Levick's insistence on methodology,
objectivity, and accuracy, marks him as a field biologist well ahead of
his time.

His scientific rigor might well be seen as a trait instilled in him by
the navy. Cape Adare is one of the remotest places on Earth but, even
in such a location, the rigid hand of naval discipline is apparent. There
is the pretense, at Campbell's insistence, of an invisible line down their
small hut to separate the quarters of the enlisted men from those of
the officers. Every Saturday, the members of the Northern Party toast
wives and loved ones, and on Sundays, they hold a church service; the
six of them, the only members of its small but loyal congregation. The
orderliness of the men's lives inside the hut mirrors Levick's approach
to recording the lives of the penguins outside the hut.

The deliberateness of Levick's approach to science would have come
as no surprise to Wilfred Bruce, the former owner of the fountain pen.
He describes Levick as "the slowest man I've ever met." According to
Bruce and others on the Terra Nova Expedition, Levick's motto is
Festina lente: Hasten slowly.

CHAPTER THIRTEEN
THE RACE BEGINS

As Levick amasses his data slowly on the sexual and social behavior of Adelie penguins at Cape Adare, his notes, written in blue-black ink, reveal that the penguins themselves hasten quickly as they go about the business of breeding. With only a narrow window when conditions in the Antarctic are favorable for breeding, the penguins are in a race to fledge their chicks before the weather turns and their food supplies disappear. When he first arrived at Cape Adare in February 1911, near the end of the previous breeding season, Levick had witnessed the fate of those chicks whose parents attempted to breed too late: they were abandoned and left to die before they could mature well enough to fend for themselves. His dissection of an emaciated fledgling had confirmed that it had starved to death, with the rocks in its stomach having made a poor substitute for food. In the Antarctic, survival and success rely on the same things: timing and food.

It is October 20, 1911. Five men—Roald Amundsen, Oscar Wisting, Olav Bjaaland, Sverre Hassel, and Helmer Hanssen—depart from

Framheim for the South Pole, taking four sledges and fifty-two dogs with them. The race for the pole has finally begun.

They are dressed in the clothing of the Netsiliks and the dogs are harnessed in the way of the Netsiliks—things Amundsen had learned when completing his childhood dream of the Northwest Passage. Four underperforming dogs are set free and left to find their own way back to Framheim, if they can (a couple do). As the men make their way to their first depot at 80°S, they encounter a very thick fog, yet Amundsen is ecstatic. They are able to navigate and find the depot easily using the flags they had set out before the winter. It is "a brilliant test," he notes, of their meticulous precautions and preparations.

At their depot they have food aplenty, as it had originally been set up with an eight-man expedition in mind. They load the sledges with the gear and food that had been left there on their first, injudicious, attempt to get away over a month earlier. Now that the race has really begun and the conditions for the Norwegians seem good, they are, as Amundsen records on October 24, 1911, "enjoying life."

It is October 24, 1911. Just as Amundsen and his four companions are "enjoying life" at their 80°S depot, Scott's complex plan for getting to the South Pole, which is nine hundred miles away from their base at Cape Evans, at last, gets underway.

Two motor sledges and four men set off across the frozen sea ice, with each sledge pulling a commendable one and a half tons of supplies slowly, very slowly. For that reason, the main party, consisting of men and ponies, is to leave a week later.

On the morning of November 1, 1911, a line of eight ponies, each pulling a loaded sledge and each accompanied by a man, sets out across the sea ice heading for the former Discovery Expedition's hut at Hut Point. A line of dark gray figures, they disappear into the nothingness and grayness of an Antarctic day that threatens blizzards and snow. At Hut Point, they pick up another two ponies and their attendants, who had gone ahead on account of the two ponies being so slow.

Five days after leaving Hut Point, the caravan of ten ponies and ten men comes across the two motor sledges, broken down and abandoned. Of the two, the one to get furthest had managed to cover a mere fifty miles from Cape Evans. The four men who had been with the motor sledges had depoted some of the food and fuel, then proceeded to carry on with the rest, pulling the sledges themselves with nearly four hundred miles to the base of the Beardmore Glacier still ahead of them.

From the outset, Lawrence "Titus" Oates, the man in charge of the ponies, has complained that the ponies are not really up to the task:

> *A more unpromising lot of ponies to start a journey such as ours*
> *it would be almost impossible to conceive.*

Indeed, the ponies are proving not the most ideal means of transport in the blizzards and snow that now confront them on the barrier. Their feet sink into the snow, forcing Scott to switch to "night" marching when, even though it remains daylight, the temperatures are lower and the surface of the snow is firmer. Icicles form on the ponies' eyelashes and occlude their vision. When they stop, the men must build snow walls to help protect the ponies from the wind because they do not suffer the cold well. Oates writes in his diary, with what seems like a measure of satisfaction:

> *Scott realizes now what awful cripples our ponies are and car-*
> *ries a face like a tired old sea boat in consequence.*

A week after leaving, the main party becomes tent-bound, stalled by a blizzard too awful for the ponies to be harnessed to their sledges and expected to walk into the wind-driven snow with its wicked windchill. Yet a third element of Scott's complex plan that involves a combination of different means of travel and supporting parties waltzes into the camp seemingly untroubled by the conditions. It is the two dog teams with their drivers Meares and Demitri Gerov. They had left Cape Evans last and were not expected to rendezvous with the main party and the motor sledges until 80°30′S, somewhat beyond the One Ton Depot. Yet, here they are now, a week ahead of schedule, within one hundred miles of Cape Evans and miles before One Ton Depot. The dogs have traveled quickly and, when at rest, lie under the snow drift untroubled by the blizzard conditions. The unvoiced truth, laid bare for all to see, is that Amundsen has chosen the best means of travel. Apsley Cherry-Garrard, who is sharing a tent with Scott, notes in his diary that even Scott thinks "Amundsen with his dogs may be doing much better."

Apart from the undeniable differences in the way the dogs and ponies travel and withstand the Antarctic conditions, another lies in what they eat. Whereas the dogs can be fed on locally available seal meat, and even their own companions, the nearest food available for the ponies must come from New Zealand, 2,500 miles away, and it is bulky. In an effort to cut down on the bulk of forage needed to be carried with them on the sledges, Scott's party have supplemented the ponies' feed with oats and oilcake, but these are not adequate nutritionally. All the ponies are losing condition. Scott is forced to start sacrificing them and to feed their meat to the dogs. Originally, he had intended that the dogs would come only about halfway to the Beardmore Glacier, but now he has no choice but to take the dogs much further. At the same time, Scott has abandoned any pretense that they might be able to take the ponies onto the Beardmore Glacier itself.

It is October 24, 1911. As Amundsen's men and dogs sit feasting on the food at their first depot on their expedition to get to the South Pole, and the motor sledges of Scott set out from Cape Evans to begin the Englishman's attempt to be first to the same geographic pole, Murray Levick is stuck at Cape Adare and going nowhere. Instead, he is left observing the Adelie penguins and their race to breed successfully in the short Antarctic summer.

It is snowing and a cold wind is blowing hard from the southeast. Clouds cover the sun and the light levels are low. Levick notes that the birds lie flat on their nests, heads facing away from the wind, and that penguins throughout the colony have become "noticeably subdued," with very little activity at all. By the afternoon, the wind drops, the cloud lifts, and the light levels improve dramatically, with the consequence that the penguins resume their frantic "love making, fighting, and building" of their nests.

Levick focuses on the fighting he observes within the colony, which he ascribes pretty much exclusively to competition between the males.

> *. . . the roar of battles & thuds of blows can be heard over the entire rookery, and of the hundreds of such fights I have witnessed, all have plainly had their cause in rivalry over the hens.*

Despite his admonition to the other men of the Northern Party that they must be certain of anything they record as fact, Levick assumes that if two birds are fighting, they must be males, or "cocks," as he insists on calling them. Indeed, an entry in his Zoological Notes says as much:

> *I conclude when I see two birds fight with flippers alone, that they are cocks.*

Even those birds fighting with beaks he believes to be males competing with each other for the "hens." He cites an instance where he sees a penguin with its eye "put out" by another's beak, leaving the right side of its face covered with blood.

Murray Levick's account of the penguins, as captured in his book *Antarctic Penguins*, with the exception of a little bit of argy-bargy among the males, describes a routine kind of domesticity:

> *. . . it was not unusual to see a strange cock paying court to a mated hen in the absence of her husband until he returned to drive away the interloper, but I do not think that this ever*

occurred after the eggs had come and the regular family life
begun . . .

While male Adelie penguins tend to be slightly larger than females, there is much overlap in their sizes, and this is further complicated by the dramatic fluctuation in their weight that occurs depending upon the amount of time they have been onshore and fasting. Levick's mistake, a prejudice born of his times, is to assume that whenever two animals are fighting, it is likely they are males and that it is the females that incubate the eggs first. As my intensive observations would later show, he is wrong on both counts.

While Levick is not able accurately to determine the sex of the birds he is observing, he can readily distinguish new arrivals to the colony. Their breasts are sparkling white and clean, while the breasts of birds that have been in the colony for a few days or more are covered in reddish-brown guano stains from lying on the ground. This allows him to make one of his most startling observations, which he records in his Zoological Notes.

> *Several times I have seen fresh cocks making love to mated hens*
> *who have shown no evasion to them until the mated cock has*
> *suddenly turned up and fought the interloper.*

It is the first intimations of the mate switching I will observe over seven decades later; it is the beginnings of his realization that the sexual behavior of Adelie penguins is nothing at all like what a Victorian gentleman might have supposed it to be.

Two days later, Levick makes another observation in his blue notebook: three male penguins newly arrived at the colony, as evidenced by their "spotlessly clean" white breasts, are near a female. Her sex is unequivocally evident from the dirty tread marks down her back left by the "love making" from a previous partner, marking her, in Levick's estimation, as "unquestionably an old arrival and a bride long past her virginity." One of these males approaches the female and is initially rebuffed. After a bit of a brawl with the other two new arrivals, he tries again and, despite the protestations of the female as she pecks him, he persists in lying next to her until he is at last accepted by her.

This too fits with my mate-switching observations and runs counter to the perceived wisdom that persisted in the penguin world for so long after Levick: that penguins are monogamous and mate for life.

That same day, October 27, 1911, Cape Adare is hit by a frightful storm. This time the activity in the colony virtually stops, and the wind is so strong that the birds face into it, allowing the wind to pass over their streamlined and interlocking feathers, leaving an undisturbed layer of air trapped next to their skin to provide some insulation; a sort of biological equivalent of double glazing.

It is at this time that Levick notices a penguin that has collected two pieces of sharply edged white quartz stone for its nest, rather than the rounded black basalt pebbles usually found on the beach. While Levick watches, a neighbor "put out its beak and stole one of the pieces!" Campbell is witness to the incident too. In a sign that the penguin study is assuming greater importance for Levick than the exploration that was their original goal, he laments, "Unfortunately I am going away sledging for four days . . ." Consequently, he asks Campbell to check on whether the quartz pieces continue to be transferred to other nests by this penguin pilfering.

It is a brilliant piece of scientific insight, a look into the mind of a natural experimenter. It will take decades before anyone thinks to do likewise: to paint pebbles and record how they are stolen and passed from nest to nest in the colony. It turns out that the stones typically move from the nests on the outside of a subcolony to the central ones. The central nests are better protected from the skuas and benefit from the largesse that their neighboring nests provide, in terms of stones that can be pilfered given the right opportunity. As a result, central nests tend to be occupied by older, experienced breeders, and their nests are often the most elaborate in terms of the number of stones lining their floor and walls.

These stones are not just for decoration. Elsewhere, Levick records how meltwater from the snow swamps nests and destroys the eggs. On one occasion, he and Browning find what he assumes is a female sitting on eggs in a nest with no stones. In fact, it is likely to be a male given that this is the first incubation spell—but no matter. The point is that the penguin is trying to sit on eggs "amidst a slush of melting snow, so that the eggs were nearly floating in water." Browning and Levick add stones to the nest that they take from neighbors and then replace the eggs, lifting them out of the water, which the bird then resumes incubating. It says something again about Levick's heart and also, in particular, the importance and value of stones to the penguins: they are their currency.

When Levick leaves Cape Adare, he travels with Browning back to Duke of York Island, arriving on October 31, where he is surprised to find 1,500–2,000 penguins nesting in the valley inland from Crescent Bay; the one where he had found the carcass of a penguin tied to a rock by one of Borchgrevink's party. No eggs can be found, though he looks carefully with Browning. To Levick, "the most striking fact about this rookery seemed to me to be the absence of open water for many miles, and the scant likelihood of there being any for at least another month or perhaps more." It is evident to him that these birds must be prepared to face a long fast during the first part of their breeding season.

Levick returns to Cape Adare on November 4, a day after Campbell has recorded the appearance of the first penguin egg in the Cape Adare colony. Either Campbell did not make any observations of the quartz stones or he saw nothing of consequence during Levick's absence, as there is no further record of them.

While Levick writes occasional references in his notes to the marked pairs in nests 1–6 in Group A and the nest containing the injured bird, the observations start to become less frequent, and by November 20, he laments that "my photography (chiefly in work with Priestley) is taking a great amount of time that I have little left for other work."

Presumably this is why Levick never really nails the nest relief patterns of the Adelie penguins. He notes that "Mated couples appear to fast absolutely until the first egg is laid, after which they go off to feed by turns." But without systematic and, importantly, frequent observations of his marked pairs, he seems not to get that it is the female that departs for sea first after both eggs are laid and that then she is away for about two weeks.

Ironically, Priestly has been doing this for him to some extent. When Priestley takes his regular meteorological observations, he monitors the nest attendance of a pair nesting near the meteorological station, noting when the sitting bird is relieved by its mate. The only problem with this—apart from its singular sample size—is that the sexes of the birds are unknown and, like Levick (or more likely upon Levick's advice), Priestley assumes that the bird taking the initial incubation spell is the female.

Yet, that aside, when I pull out my 1914 battered green-covered copy of *Antarctic Penguins* in my cabin on the *Shokalskiy*, there on pages 91–93 are Priestley's records of the nest attendance. They show the first incubation spell was thirteen days. This would almost certainly have been the male sitting on the egg while the female was away at sea feeding. Then the female returned and took the second incubation spell, which also lasted thirteen days. Once the female was relieved, she went away for four days and, thereafter, coinciding with the hatching of their chick, they alternated nest attendance every day or two. It is about as typical a nest relief pattern as one can get for Adelie penguins. Yet Levick does not seem to grasp its significance.

Instead, in his book, he falls back to the quaint writing that I had found of such limited value when I had first taken it with me to Cape Bird in 1977. He writes, "the couples took turn and turn about on the nest, one remaining to guard and incubate while the other went off to the water." But he then gives in to the kind of anthropomorphism that is completely at odds with the scientific version of Levick that I have started to get to know, such as when he describes birds returning from the first feeding trip to sea and stopping to linger with other penguins on the ice-covered beach. While he erroneously assumes that they are males, I haven't got a problem with that, but I do with the way he expresses himself:

. . . they were sociable animals, glad to meet one another, and, like many men, pleased with the excuse to forget for a while their duties at home, where their mates were waiting to be relieved for their own spell off the nests.

Where, oh where, has Levick the scientist gone, the man who marked nests and individual penguins, the man with the means to uncover even more than he did? It seems that during this period, Levick the Photographer has trumped Levick the Scientist, and so he spends a lot of time down at the ice edge, photographing penguins leaping into the water or out of it, as well as penguins porpoising. Even with modern digital cameras with their autofocus systems and autoexposure, such actions are not easy to capture. Given his equipment and inexperience, Levick's photographs are quite remarkable in that regard. However, they could only have been achieved by spending time doing that at the expense of time spent monitoring his nests.

A few days later, on November 24, 1911, Levick records that the skuas "are stealing a large number of eggs." However, six days later, he notes that "a large number of nests in the rookery are now to be seen deserted." At first, Levick puts this down to the skuas taking the eggs and mostly wonders what has become of the parent birds. Although he goes on to speculate that while the eggs in some nests "may have first been filched by skuas, and the nest then deserted," some accident to the eggs, which caused them to be exposed, could produce the same result. At any rate, he emphasizes that "the number of deserted nests is now very great indeed."

Four more days and on December 4, 1911, he states that "The number of deserted nests continues to increase." Furthermore, according to him, "A large number of desertions seem to be due to the eggs getting rolled out of the nests accidentally, as many are to be found on the ground, frozen, which have not yet been eaten by skuas." In reality, he has missed the boat again. The timing of these desertions corresponds to birds left fasting on the nest and being unrelieved by their partners. Eventually they desert and that is why Levick finds the unattended eggs lying within the subcolonies, presumably knocked out of their nests by the movements of the colony's dwellers rather than fighting between the males, which seems to be Levick's fallback position every time he suspects aggression.

I daresay the cocks are the greater offenders in this respect . . .

Levick had all the right ingredients in his study to have gotten this. He had marked birds in marked nests. He had only to monitor them daily to be able to pick up the pattern, the nest relief pattern, noting which bird was present and for how long. But perhaps he was being my Eivind Astrup rather than my Sherpa? Perhaps he was showing me a means of travel that I could use later rather than his doing the traveling before me?

CHAPTER FOURTEEN
COMPETITION

Roald Amundsen has certainly learned the lessons from Eivind Astrup. Using skis and dogs, his party of five leaves their first depot at 80°S on October 26, with four sledges, each carrying 880 pounds, twice the weight of Scott's. Although the ice conditions and weather across the barrier are oftentimes difficult; with fog and snow drift making visibility difficult; and gales, crevasses, and sticky ice making progress difficult; they still make excellent time. Amundsen knows the importance of resting both dogs and men in this environment. He sets specific targets for each day that are mostly about fifteen miles and never more than twenty. They cover the distance in five to six hours, with the dogs pulling both the sledges and the men on their skis.

By contrast, Scott and his men plod through the snow with their ponies, moving at only about half the speed of the Norwegians and sometimes much less. Moreover, their days are long: marching for eight to ten hours to attain distances of ten miles or so, and rarely more than thirteen. By the time they stop each day, the ponies and men are completely exhausted.

That is not to say that Amundsen has it easy. He is just better prepared for the conditions. They are well equipped with the "eskimo" style of their clothes. In fact, the garments made of reindeer fur, so helpful when laying the depots in the brutal cold before winter, are now too warm. They leave them unworn on the sledges, finally abandoning them altogether to save weight. They have the tools and experience needed for navigating. The men are all expert skiers and dog drivers and, importantly, they are used to these types of conditions from their northern upbringing. Amundsen and his party experience more than twice as many days of gale force winds on their way to the Pole as does Scott's party, but they still manage to travel on more than half of them. Scott, his men, and their "unpromising lot of ponies" are unable to travel on any of them.

One thing that distinguishes Amundsen's journey from that of Scott is that Scott is following a known route: the "road" up the Beardmore Glacier discovered and traversed by Shackleton in 1908, that he knows for certain can take him from the barrier to the Polar Plateau. Amundsen is heading south into uncharted and untraveled lands. His means of getting from barrier to plateau is completely unknown, if one exists at all.

Amundsen has been heading due south, happy to stay on the barrier ice as long as possible. But on November 17, they have gone as far south as they can on the barrier and they find themselves at the base of the Transantarctic Mountains, a line of unbroken peaks of twelve thousand feet, fifteen thousand feet, or more that stretches for 2,000 miles from Cape Adare across the whole of the Antarctic continent. Their only option to get to the pole is to somehow find a way to travel up the mountains to the plateau, which consists of ice that is miles thick and is held behind the mountains like a dam, with the ice spilling over between the peaks as heavily crevassed, steep, and dangerous glaciers.

No obvious way beckons. Setting up camp, Amundsen and his men scout options. Amundsen chooses an especially daunting one: a steep, wide glacier marked by "crevasses and chasms," which he will eventually name Axel Heiberg Glacier after his patron for both his Gjoa and Fram Expeditions.

The route is so steep that in places the dogs must crawl on their bellies, struggling to get the purchase they need to pull themselves up, let alone the sledges with them. In the steepest pitches, they double-team the dogs, using two dog teams per sledge and relaying: going back to repeat the climb with the sledges that had been left behind. All the while, the dogs are egged on by their drivers and, particularly Bjaaland, the most brilliant of skiers, who goes ahead to coax the dogs up what Amundsen describes as "pit after pit, crevasse after crevasse, and huge ice blocks scattered helter skelter."

It remains one of the greatest of polar feats ever achieved: in four days they manage to negotiate an unknown and deceptively dangerous route, pulling one ton of food and equipment up ten thousand feet over a distance of forty-four miles. Yet, through all the journey so far, Amundsen never loses his connectedness to the world around him. He never thinks of it as an awful place. To the contrary, he is smitten with its beauty as much as he had been with Sigrid Castberg, and he is just as glad to be here.

> *Glittering white, shining blue, raven black . . . the land looks like a fairytale. Pinnacle after Pinnacle, peak after peak—crevassed, wild as any land on our globe, it lies, unseen and untrodden. It is a wonderful feeling to travel along it.*

Once on the edge of the Polar Plateau, they set up camp, and as per Amundsen's original plan, they shoot all but eighteen of the dogs who are then fed the flesh of their valiant but not so lucky comrades. At Amundsen's insistence, the men eat some of their dogs' flesh too, in order to ward of scurvy, his lesson from the *Belgica* and Frederick Cook still etched in his memory.

Despite Amundsen's determination to use dogs with the consequence of that being the need to sacrifice the majority of them, it is not something that he relishes.

> *. . . there was depression and sadness in the air—we had grown so fond of our dogs. The place was named the "Butcher's Shop."*

Levick is at Cape Adare and continuing to make notes about the penguins in his blue-bound book. His own revulsion to what he sees is, unlike Amundsen, not apparent from what he writes but, rather, what he does with what he has written.

After writing down his initial observations of mate switching, mysteriously, at some time afterward, he goes back and covers up the next sentence with a pasted piece of paper and Greek letters.

Hove to off Cape Adare, with the penguins in sight of me and the shattered remains of the hut where Levick wrote his Zoological Notes, I sit in my cabin on the *Shokalskiy* reading once more through his Zoological Notes on my laptop, using the photos I had taken earlier in the book-lined apartment in London.

No matter how many times I return to them, I am always shocked by the first evidence of his strange behavior: a piece of paper cut out to cover a few lines of text and covered in Greek symbols.

This was clearly an afterthought: something he decided to do after writing his initial observations. In the first instance, on October 17, the paper is cut so that it covers only the second half of one line before covering up the six subsequent lines of text completely.

The paper that covers the text is the same paper as the journal itself and presumably has been cut from elsewhere in the journal or one just like it. Indeed, one of the pages in the notebook near the front has been ripped out. At 300 percent magnification, the patched paper and the underlying paper can be seen to share the same embossed details. The covering paper is also the same weight and color as the original.

The covered-up lines are faintly visible beneath the pasted paper. By boosting the magnification and contrast, I can make out the text below. He had crossed it out by running a squiggly line through it before covering it up. Such extraordinary secretiveness for what are supposed to be just scientific notes.

The code itself is easy to break: a schoolboy's code, no doubt learned by the young Murray Levick at St Paul's School in London, an elite public school for boys, as a way of passing "secrets" to his classmates.

I decipher the coded section after his initial observation on October 25 about mate switching. It reveals that the mated pairs responded to the mate switch by copulating frequently afterward. It is such a shame that he chose to keep such observations secret because, remarkably, they are eighty years ahead of their time.

———

It is October 30, 1993. I am at Cape Bird observing the courtship period of the Adelie penguins. This time I am accompanied by a softly spoken postdoctoral fellow, Fiona Hunter, who has a toughness and ease in this environment that even Amundsen would admire.

When Adelie penguins switch partners during the courtship period, as I had observed in my earlier research, a potential dilemma for the male penguin arises if his female partner already has sperm inside her reproductive tract from another male. The costs of that are potentially very high.

Penguin eggs and chicks require a lot of care and investment by both parents. It is impossible for the female to rear the eggs and chicks alone. However, if a male were to spend an entire summer rearing another male's offspring, his own evolutionary fitness would be mark-edly reduced while benefitting that of the female's initial lover.

Fiona and I discover that male Adelie penguins have developed counterstrategies to use in exactly those situations where their partners have already been doing the wild thing with another male: they bonk like crazy. Like once every three hours. And whereas pairs typically put a halt to conjugal capers once the first egg is laid, as there is plenty of viable sperm remaining in her reproductive tract to fertilize the second egg just before it is laid three days later, males with partners that switch to them from another male continue to copulate right up until the time the second egg is laid.

These males are engaging in a kind of biological warfare using sperm as their weapons. Their frequent fornication is in an attempt to swamp any sperm from their predecessor that might be remaining in the female's reproductive tract with their own sperm.

This sperm competition, where the sperm of two-timed males are in a race to beat the other penguin's sperm to fertilize the female's two eggs is, I suppose, not much different to having two parties racing to the South Pole. There can be only one winner and much will depend on the strategies they employ to get to their goal. DNA evidence indicates that this counterstrategy by the two-timed male penguins, whereby they essentially blast the female's cloaca with sperm, is effective: very few male Adelie penguins end up rearing offspring that are not their own.

In other species of penguins I have studied, such as Erect-crested penguins, mate-switching is not so common and copulation rates are an order of magnitude lower. Or, put another way, a more manageable, if less enjoyable, once every thirty hours. In fact, as part of their Sperm Wars, Adelie penguins fornicate so often during the courtship period that the males can literally run out of sperm, and, even then, they will continue to bonk their female partner even though they are firing blanks.

Fiona and I know all this because we collect the semen by swabbing the females' cloacas after they have been mated, and with the most bizarre piece of methodology I have ever used in my scientific career: by inducing male penguins to copulate with a dead penguin. We had a taxidermist mount a dead penguin in such a way that it resembled the mating position of a female lying in the nest: tail raised, head tilted back. To collect sperm from males we wrap a cellophane covering over the dead bird's nether regions and simply collect the semen deposited on it by the males. The males need no encouragement at all to mate with the dead female.

In fact, so little does the "female" need to resemble reality that we discover we can use a fluffy toy penguin instead. Fiona had bought one from the Antarctic Centre in Christchurch just before our flight to Antarctica. When we place the toy penguin prostrate on the ground beside a subcolony of penguins, the males practically line up to mate with it.

This accords with another of Levick's coded observations. On November 10, 1911, there is a large passage pasted over. Transcribing the code of Greek letters, it reads:

> *This afternoon I saw a most extraordinary site (sic). A Penguin was actually engaged in sodomy upon the body of a dead white throated bird of its own species. The act occurred a full minute, the position taken up by the cock differing in no respect from that of ordinary copulation, and the whole act was gone through down to the final depression of the cloaca . . .*

The dead bird is the carcass of a fledgling from the previous breeding season, one of those Levick had observed the previous February that had been too slow to fledge. As such, it is more an example of necrophilia than it ever was sodomy.

The pasted text continues, although, perversely, Levick chooses to use English instead of Greek:

> *On returning to the hut I told Browning, hardly expecting to be believed, but to my surprise he at once said that he had seen the same thing several times, done to dead bodies . . .*

Then, just as perversely, Levick switches back to using his code of Greek letters to say where these incidents observed by Browning have occurred.

It turns out that such necrophilia, as Browning and Levick's observations might suggest, is not all that uncommon in Adelie penguins. Neither is Levick so far off the mark by implying that such behavior is similar in cause to the homosexual behavior that he, and subsequently I, would see. Our observations of homosexuality and necrophilia—indeed, my research with Fiona that showed male penguins will just as readily get it off with a fluffy toy as a female penguin—stem from the same root cause: males are not very discriminating in what they will fornicate with because the costs of making a mistake are so low. Sperm are cheap.

It all comes down to the relative size of the contribution that males and females make to mating. A male passes on his genetic material by way of sperm: tiny, almost invisible tadpole-like things that he produces in the hundreds of millions with every ejaculate. Sperm are cheap to produce and misplacing or misusing even a few million of them is of no consequence: there's plenty more where they came from, for the most part.

By contrast, the female Adelie penguin passes on her genetic material in the form of eggs, and she produces just two of them each year. They are large and contain yolk and albumen to nourish a developing embryo. The females have a lot invested in each egg and they need to be cautious about with whom and where they mate: get it wrong and the whole breeding season's opportunity is lost.

So, simply as a consequence of the differences in the size of their respective investments in eggs or sperm, females and males behave differently when it comes to sex. It pays for females to be extremely choosy about who, when, and where they mate. Males, on the other hand, can afford to throw caution to the wind and sow their wild oats wherever they like in the hopes that some will germinate and bear fruit, so to speak. It's a condition that affects males of many species, including humans.

Levick more or less hints at this when he notes that "now the season is so far advanced there must be a certain number of both cocks and hens wandering about who have been left out in the race for partners . . ." He thereafter switches to more pasted Greek letters to excuse the behavior of the males on the basis that it is the only option left to them.

Who would have thought that a race to breed could be used as an excuse for necrophilia or sodomy; just as who would have thought that a race to the South Pole could lead to so much bloodshed?

It is December 9, 1911. Scott's party has finally made it to the base of the Beardmore Glacier, trudging through storm after storm after storm. The five ponies that remain are thin and tired, at times sinking in the snow up to their bellies. Oates walks them away and shoots each in turn. The flesh is hacked from their bones. There is blood and guts everywhere. The men name this blood-stained place Shambles Camp. In that regard, they are not so different from the Norwegians: "shambles" originally designated a butcher's place for slaughtering animals. Even its more modern connotation as a place of carnage seems appropriate.

The pony meat is frozen in the snow to act as a depot for the returning parties of men. Even at this stage, some of the men are looking almost as gaunt and tired as the ponies. None more so than the four men who had been in charge of the motor sledges: they have man-hauled their sledges through the same awful conditions as the ponies for almost four hundred miles to get this far.

It is a strange contradiction, indeed, that Scott can be so against sacrificing dogs for the purposes of Antarctic travel, yet he is so prepared to sacrifice the ponies, which Cherry-Garrard describes as "a horrid business." Unlike dogs, which might be fed to their companions and thereby allow at least some of them to make the return journey, from the outset there was never any way to carry enough food to get any of the ponies back to base, even were they capable of walking another foot, let alone another four hundred miles. Scott's plan had always called for the ponies to be killed.

By now Scott's party is behind schedule. They have already started to eat into the rations intended for use on the Polar Plateau at the top of the glacier. The dogs should have turned back by now, but Scott makes a last-minute decision that they will go on for a couple more days, even though that means all the men giving up one biscuit per day of what are already inadequate rations for such backbreaking, calorie-sapping work.

Scott's ponies all slaughtered by Shambles Camp; more than half of Amundsen's dogs slaughtered by the Butcher's Shop. However, nothing emphasizes the differences between these two expeditions more so than the locations of these abattoirs. Amundsen was already on the Polar Plateau and only 274 miles from the pole. Scott is still on the barrier, still with 10,000 feet to climb and still more than 430 miles to go to the South Pole. Plus, the dogs had been slaughtered on November 21, eighteen days before the carnage of Shambles Camp.

By any measure, Amundsen is well ahead, and his dogs and his men have proven their worth.

It is December 9, 1911. While Scott and his party are at Shambles Camp with the blood and entrails from their five ponies spread about the snow, Amundsen and his four men stay in bed late, resting in their sleeping bags. The day before they have passed Shackleton's furthest south record, and they are now within ninety-five miles of their quarry. Amundsen has declared it a rest day "to prepare for the final onslaught."

They make their final depot, taking care to mark the whereabouts of the cairn containing around two hundred pounds of food and fuel with thirty black planks taken from empty sledging cases in a line three miles long on either side of the depot. The next morning, the party that now consists of the five men, three sledges, and seventeen dogs, heads off south in glorious sunshine. "Sledges and ski glide easily and pleasantly," according to Amundsen, across what is now a "quite even and flat" surface.

The competition is almost over.

CHAPTER FIFTEEN

TIMING

E ven when not part of the competition, it is possible to be affected
by it.

December 11, 1911. Douglas Mawson, aboard his ship the SY
Aurora, arrives at Macquarie Island, a large subantarctic island 960
miles south of Hobart. His Australasian Antarctic Expedition had
finally got underway from the Tasmanian capital nine days earlier.
Unable to go to Cape Adare as planned, because Scott has installed
Campbell's party there, Mawson is forced to head south into the
unknown and unexplored parts of the big white continent directly
below Australia. First, he is stopping at Macquarie Island to erect radio
towers and leave a party there in the hopes that they may be able to relay
radio signals between his base in Antarctica and Australia. He had left
Paquita in Adelaide six weeks earlier, and though their parting had been
more sweet than sorrowful, radio contact with Australia would be a
bonus for his fiancée, not to mention all the other benefits it promised.

Macquarie Island is surprisingly steep and rugged, like a piece of
land that in another life had been a shark's tooth: a high craggy ridge-
line extends down the entire backbone of the island, making access

to its heights difficult. At its northern end, however, there is a small 350-foot-high headland where they will build their wireless communication station.

The black shingle beach is dotted with elephant seals, some alive, some dead. And among the four types of penguins to be found on the island, there are large congregations of the beautiful orange-cheeked King penguins, the very ones that Johann Reinhold Forster, in a case of mistaken identity, had ascribed to the first Emperor penguin seen in Antarctic waters. Despite their numbers, jammed onto the shore between the sea and the steep hillside, there are far fewer penguins than there should have been. For in their midst are large steam digesters that had been set up by a New Zealander, Joseph Hatch, a dozen years earlier. These are used to render down the penguins so as to extract oil from them. Other pots are used to boil up the blubber of seals to also extract oil, the unwanted parts of their bodies left to rot upon the beach.

Mawson is appalled. Yet Hatch's ship, the *Clyde*, which had been preparing to take the penguin and seal oil off the island, had broken its moorings in a gale and been wrecked. Mawson has another ship, the *Toroa*, coming to Macquarie Island to bring supplies. He offers to take Hatch's stranded men back to Australia on the *Toroa*, but only on the condition that it also takes back the penguin and seal oil that can be salvaged and that money derived from selling the oil goes to his expedition.

Mawson is happy to leave all the death and destruction on Macquarie behind as the *Aurora* finally heads south, but he will not forget the travesty he has witnessed. In the years that Hatch's men operate the penguin digesters on Macquarie Island, they kill and boil around three million penguins, almost all of them King penguins.

<hr />

It is December 14, 1911. This is the twenty-first day in a row that Murray Levick has not made an entry in his personal diary. It is as if he does not want to make a public record of what he is seeing. While there are no penguin digesters to appall him, he is repulsed by the behavior of what he calls the "hooligan cocks."

In his Zoological Notes, Levick had recorded that the first chicks hatched on December 7, although, as one nest had two chicks and within a brood they usually hatch at least a day apart, the first was probably hatched on December 6. He also noted that, "owing to the wind the old birds are sitting very closely and there are probably many hatched already." Given that the first eggs were seen in the colony by Campbell on November 3, this would be consistent with an incubation period of about thirty-three days, which is pretty normal for Adelie penguins.

Levick correctly deduces the limitations on the parents now that their chicks have hatched:

> *Whilst the chicks are small the two parents manage to keep them fed without much difficulty; but as one of them has always to remain at the nest to keep the chicks warm, guard them from skuas and hooligan cocks, and prevent them from straying, only one is free to go for food.*

In a sign once more of his scientific approach to his study, Levick weighs chicks at different ages and finds that they grow remarkably quickly on their diet of krill, which is brought to them by their parents and then vomited into their gaping and begging mouths. The largesse of the Southern Ocean during this brief period of the summer means that there is food aplenty, and the chicks are quickly transformed from slim gray dots of fluff into chicks with bulging bellies that would be the envy of any male in Newcastle upon Tyne. They amuse Levick.

> *To see an Adélie chick of a fortnight's growth trying to get itself covered by its mother is a most ludicrous sight.*

From the penguins' perspective, it is no laughing matter. The secret to their success depends upon timing. The very worst a penguin can be is late. To be successful, the parents are in a race to bring enough food to the chicks to ensure that they become large enough to fledge—large enough to be independent and survive on their own. And even then, simply fledging a chick is not enough: low-weight fledgling chicks are unlikely even to survive long enough to return to the colony to breed someday. They are destined, instead, to become Darwin's dead meat: the non-survival of the unfittest.

It is December 14, 1911, 3:00 P.M. The sky is blue and clear, the sun is bright and high overhead. Roald Amundsen is on skis, out in front of the line of three sledges and their dog teams, leading the way across a flat, white, nondescript tableau that stretches in every direction as far as he can see. No one place looks any different from the other. But, then, his men driving the dogs, who have been watching their sledge meters to measure the distance traveled with all the intensity of skuas eyeing an exposed penguin egg, shout in unison, "Halt!" They are there. This is the South Pole. And, what is as clear as the day: they are the first humans to ever be here. There is no sign of the Englishmen.

The five men, who have accomplished so much, quietly shake each other's hands. There is no cheering or back slapping, no hugging or wild exultation. In a measure of Amundsen's leadership, at his insistence, all five men grip the post with their "weather-beaten, frostbitten fists" and plant the Norwegian flag to mark the geographic South Pole, with Amundsen naming this expanse of nothingness King Haakon VII Plateau.

Amundsen, highly sensitive to the criticism directed at both Cook and Peary, then sets out to make sure there could be no doubt that they have, indeed, attained the South Pole. The next day, in bright sunny conditions, Amundsen and one of the men use their sextant to take accurate readings of the path of the sun throughout the day. To make sure that they have boxed in the area containing the exact position of the pole, he sends the other three men out skiing for ten miles at 90° intervals from the direction where they have come. Each carries a twelve-foot-long spare sledge runner, which at the designated distance (determined by their time of travel), they plant in the snow. Atop the sledge runner is a black flag and a small bag containing a note for Scott.

Amundsen's measurements and calculations show that they had stopped five and a half miles from the exact position of the pole. On December 17, they move camp to a new position and all four navigators in the party participate in taking readings over the next twenty-four hours to be doubly sure of their position, countersigning each other's navigation books, in contrast to the lackadaisical records of Peary. In a suggestion that Amundsen does not quite believe either Cook or Peary's claims of having gotten to the North Pole, he writes:

> *It is quite interesting, to see the sun wander round the heavens at so to speak the same altitude day and night. I think somehow we are the first to see this curious sight.*

In fact, Amundsen's subdued reaction upon reaching the South Pole betrays that his real ambition had been the North Pole instead.

> *I have never known any man to be placed in such a diametrically opposite position to the goal of his desires as I was at that moment. The regions around the North Pole—well, yes, the North Pole itself—had attracted me from childhood, and here I was at the South Pole. Can anything more topsy-turvy be imagined?*

Their measurements show that they are camped now within 2,500 yards of the actual position of the Pole. Amundsen sends his men out with pennants to mark the area a few miles in each direction to make triply sure they could not be denied their priority of getting to the South Pole, whether that is the prize he had originally yearned for or not.

It is December 18, 1911, when Amundsen and his men erect their reserve tent to which they attach a bamboo pole with the Norwegian flag. Poleheim. Inside they leave some equipment they no longer need and a letter to King Haakon VII with a covering letter addressed to Scott. Outside they leave one of their sledges. Upon their arrival at the pole three days earlier, they had killed one of the dogs and fed it to its companions. Now they return northward, bound for Framheim in the Bay of Whales; five men on their skis, sixteen dogs, and two sledges.

All has gone to plan. Has there been luck? Certainly. For the most part, though, they have made their own luck through Amundsen's meticulous planning and their adherence to traveling by Eivind Astrup's method of using skis and dogs. They are all as comfortable on skis as they are experts at driving the dogs and sledges. They have never been in any danger of not having enough food or fuel. Amundsen has depoted and carried three times as much food per man as has Scott. Amundsen's margins for error, like those of a penguin with a fat belly, are much greater than Scott's. Scott must depend upon good luck and good weather. Amundsen does not.

It is December 14, 1911, the day that, unbeknownst to Scott and his men, Amundsen has arrived at the South Pole. They are more than four hundred miles behind. The twelve men are hauling three sledges carrying two hundred pounds of weight per man up the 120-mile-long Beardmore Glacier, climbing ten thousand feet to the Polar Plateau. The ponies had been put down five days earlier. The dogs, along with their drivers, Meares and Demitri, had turned back to Cape Evans two days after that.

The snow is deep. None of the men are proficient enough on skis to use them as profitably as the Norwegians might have done, even without their dogs. Scott writes, with some exasperation, in his diary the day the dogs leave:

> *Ski are the thing, and here are my tiresome fellow-countrymen*
> *too prejudiced to have prepared themselves for the event.*

Though rather than blame the men, Scott should shoulder some of it himself. He had taken the young Norwegian skiing expert Tryggve Gran with them to Antarctica at Nansen's suggestion, but then had not put a priority on using him to get the men up to speed with skiing, despite their many months of living in the snow and ice of Antarctica. Now they are on the Beardmore Glacier, when skiing is their best option for making progress. They ski when they can and trudge when they cannot, often sinking in snow up to their knees, or even thighs. Slipping and falling, they inch their way up the glacier with their heavy loads, draining valuable energy from their already stressed bodies. And, if it is not the snow that is a bother, it is, according to Bowers, "the perfect mass of crevasses into which we all continually fall; mostly one foot, but often two, and occasionally we went down altogether."

It is December 21, 1911, when they finally reach the upper glacier near the start of the Polar Plateau. It has been no four-day transition from the barrier to the plateau for them: it has taken them three times as long as Amundsen. At this stage, Scott sends back the first of the supporting parties: one of the sledges and four men, including Dr. Edward Atkinson and Apsley Cherry-Garrard. Before they leave, Cherry-Garrard notes that Scott tells Atkinson "to bring the dog-teams out to meet the Polar Party" upon their return from the pole.

The next day, the remaining two four-man sledges continue on toward the pole. One is to be the advance party that will go onto the pole, the other is the supporting party needed to take food and gear as far as possible for a final depot. As Cherry-Garrard writes:

> *The final advance to the Pole was, according to plan, to have been made by four men. We were organized in four-man units: our rations were made up of four men for a week: our tents held four men: our cookers held four mugs, four pannikins and four spoons.*

On the last day of 1911, without explanation, Scott orders that the four men on the other sledge must depot their skis. It is a baffling decision because as he himself notes, thereafter, "We started more than half an hour later on each march and caught the others easy. It's been a plod for the foot people and pretty easy going for us."

It seems clear that Scott intends to take only four men onto the pole and they are to be the four pulling his own sledge, which consists of Wilson, Titus Oates, and the seaman Taffy Evans, in addition to himself. However, two days after ordering the others to leave their skis behind, while confirming that three of the men on the other sledge are to go back, Scott adds the remaining man to the team to go onto the pole: Birdie Bowers, the five-foot, four-inch Scotsman who had already proven his mettle with Wilson and Cherry-Garrard on the winter journey to the Emperor penguins at Cape Crozier. In addition to his strength, Scott wants Bowers because of his ability as a navigator: none of the others on his sledge can use a theodolite properly. This, however, creates enormous logistical problems when sledges, tents, cooking, and food have been arranged for groups of four-men. Plus, there is the other glaring issue: Bowers has no skis.

The sensible thing might have been to have swapped Bowers for Oates. Wilson, and Atkinson before him, have warned Scott that Oates is suffering because of his feet. However, Oates is an army man and Scott is determined to have both the army and navy represented when they finally, victoriously, get to the South Pole—as now seems likely, given that they are only about 150 miles away.

It is January 4, 1912. The three-man sledge party bids farewell to the Polar Party and turns back for Cape Evans. At first, they continue to watch the five men, dark dots, four on skis and with Bowers on foot in their center, as they pull their sledge and quickly disappear into the white eternal nothingness that lies before them. The three men have a harrowing journey ahead of them, when death will seem as much a relief as it will seem inevitable, and yet, it is nothing to what lies ahead for the other five.

Scott, though, is largely oblivious to that. His diary entry for the day proclaims, "At present everything seems to be going with extraordinary smoothness."

It is January 4, 1912. Good fortune, it seems, is also with the Northern Party. Rather than being left alone, they are boarding the *Terra Nova* to be taken away from the prison that Cape Adare had become for them.

For the preceding week, pairs of the men had been taking turns to camp at the top of Cape Adare near Nicolai Hanson's grave to keep a lookout for the return of the *Terra Nova*. Yesterday morning, at 8:30 A.M., Levick had seen the little ship steaming toward them. He hoisted "the flag to signal to the hut."

Today, the captain of the *Terra Nova*, Harry Pennell, decides that they must get out of Robertson's Bay quickly before it gets closed in with pack ice. The men are forced to abandon their loading and get onboard the ship immediately. Levick leaves behind his precious samples of lake water and two leopard seal skins.

Most significantly, however, it is his penguin study that Levick is leaving behind. And this at a time when the colony is once again going through a major change. Just five days earlier, as Campbell wrote in his diary, the chicks had started to form crèches.

> *The penguin chicks are able to walk now and huddle together in batches.*

But the world's first penguin biologist must board the *Terra Nova* and leave Cape Adare before he can observe much of that.

Though he has another, much more pressing problem: what to do with his Zoological Notes? The plan is for the ship to transport the Northern Party down the coast to Terra Nova Bay so that they might, at last, carry out some real exploration. Levick will need to leave his Zoological Notes onboard the ship, but what if someone should read them and the atrocities they contain?

I am at Cape Adare and about to leave too, but first, a final look. I wait on the deck of the *Shokalskiy* for my turn to get into a Zodiac, which has been lowered down the side of the ship by crane. The sea is heaving and it is an awkward step from the runway onto the gray pontoon of the Zodiac. It is late in the evening and already the sky is gloomy: twenty-four hours of daylight, yes, but the sun is below the horizon and the light levels are low. The dark clouds do not help.

We take the Zodiac through the ice-encrusted surface of the sea to Ridley Beach. In places, the underlying black stones of the beach are exposed, but all along the shore there are pieces of ice, either marooned there or being joggled in the swell. Atop most of these pieces of ice there are white-chinned, fledging Adelie penguins, thousands of them, congregating on the shoreline, waiting to take their first swim. Their parents have stopped feeding them. They have been left to their own devices now. Their only hope of survival lies in the surging mass of gray water in front of them. They stare at it, immobile. To a casual observer, it appears as if they are frozen with fear, waiting to build up the courage. And well they might. There is no practice run. They must dive in and then they are on their own in the water for the next two years or so. They will have to learn to hunt for the krill and fish they need by themselves. There are no teachers, no penguin school. And so they wait.

We slow the Zodiac to a crawl, cruising a yard or two from the large blocks of ice that are being bumped up against the shore by the swell. Every now and then one of the fledglings, sometimes followed by a group of its classmates, will throw itself into the water, which is so full of brash ice that it is like jumping into a churning pool of ice cubes. They keep their heads up, as if afraid to put them under water, and they flap their floppy and fairly ineffectual flippers. Perhaps it not the water they are afraid of but what it contains? Not far offshore, we see the dark hump of a leopard seal cruising the shoreline also.

The lot of the chicks further inland is far worse. Those that remain on Ridley Beach, which I can see still standing among the scattered stones that are all that remains of their subcolonies, still with down covering parts of their bodies: they have no future. There are few adults around and though they beg from any that come nearby, it is not with any success. Parents feed only their own chicks and most parents have already left the colony to fatten up so they can undergo the rigors of molt, whereby they replace their worn-out feathers by growing new ones. It is like putting on a new change of clothes with better insulation before the ravages of the Antarctic winter set in.

The sea may be a scary proposition for the white-chinned chicks massed along the shore, where their survival will be uncertain, but for those that remain on land it is death that is certain. It is just as Levick had observed when arriving at this very beach.

We take the Zodiac out into the black waters of Robertson Bay. Penguins are clustered on icebergs with curved sides sculpted by the sun and tides, as if Kathleen Scott herself had a hand in the making of each. Occasionally the penguins dive from the sides of the bergs, like a brief black-and-white waterfall, as the birds respond to whatever instinctual urges are telling them to get the hell out of there. The dark brooding clouds and the shortening days speak of further darkness: the coming winter.

Even the white-chinned fledglings, as fearful of the water as they seem to be, nevertheless throw themselves into it: it is their route to getting out of there, their route to not just future success, but to a future at all. The surface of the sea is starting to freeze over, creating a carpet of ice. The fledglings make their way through it, heads up, more like ducks than penguins. Two clamber onto a small piece of pack ice in front of the Zodiac, but their respite is short.

The large bulbous head of a leopard seal glides toward them. Eyes wide with fear, the young penguins dash, helter-skelter, to the other side of the floe. The three-meter long seal propels itself completely out of the water and slides across the surface of the ice toward them. The gray-olive skin on its head and back contrasts with its white undersides, which are dotted with black spots. At the last moment, it opens its enormous jaw: a line of massive teeth and a huge pink tongue engulf the slowest of the chicks, grabbing it around its midriff.

There are no sounds. No grunts from the seal. No screams from the chicks. This is a silent assassination. The seal slips into the sea, penguin chick in its mouth. The seal dives with the penguin clamped in its jaws, a futile wave of the penguin's flipper the only visible sign of a struggle it cannot win.

Returning to the surface with the penguin's body now limp, the seal slaps the penguin on the surface of the sea with vicious slaps, pulverizing its flesh. Eventually, the leopard seal tilts its head backward and, with a big seal bite, gulps down the penguin: its lifetime since fledgling measured in hours, if not minutes.

During their time at Cape Adare, Victor Campbell and Murray Levick studied the leopard seals in the only way that their Victorian backgrounds had prepared them to do: they shot the seals. Levick dissected one and found that, in its stomach, it had the remains of eighteen penguins.

Despite the dangers that leopard seals pose to adult and fledgling penguins alike, they kill so relatively few compared to the death and destruction about to be exacted by the oncoming winter. In the Antarctic, being late is never a good option for bird or man: it leads to running out of adequate food and being battered by the coming storms.

On January 4, 1912, the future had looked so bright for both the Polar and Northern Parties. How wrong that would turn out to be for all of them. And while luck is always a factor for survival in the Antarctic—an inopportune storm here, a leopard seal there—the chief determinant of success, be it for penguins or men, is most often timing. Doing the right things at the right time.

A storm is coming. We pull anchor, and get the hell out of Cape Adare.

PART FOUR

AFTER
CAPE ADARE

Rape

Only humans commit murder; only men commit rape. Until recently, pretty much any biologist would have agreed. Nobel laureate Konrad Lorenz argued convincingly in the 1960s that behavioral feedback mechanisms like submissive postures prevent animals from deliberately killing members of their own species. Similarly, sexual perversions are seen as a peculiarly human condition. Sadomasochism is hard to imagine in birds or other beasts. Rape and gang rape are things we see in our newspapers not in the animal kingdom.

Yet over the last four decades, close inspection of creatures from ground squirrels to great apes, from white sharks to black eagles, from timid mice to the king of the beasts, shows that they all commit murder. It turns out that there are times when, biologically, it pays them to kill their own kind. Murder really can be an adaptive strategy. Murder really can be programmed into an animal's DNA. Who would have thought?

So if animals can be programmed to commit murder, does it take too much of a stretch of the imagination to think that they could also be programmed to coerce other animals into having sex: to rape? Rape is such a heinous crime that nowhere in our society do we allow for the possibility that it may represent a personality trait rather than a personality defect, a biologically determined characteristic like the color of our eyes. We might not forgive rape, but perhaps evolution could? Perhaps conditions could exist whereby behaviors that give an animal an advantage in producing offspring, as part of natural selection's own competition, might lead to rape?

One thing we know for sure is that for the Victorian-bred Levick, such thoughts would never have crossed his mind until he went to Antarctica and was more or less forced to observe its penguins.

CHAPTER SIXTEEN

HOOLIGANS

I n my cabin on the *Shokalskiy*, with the penguins and the shattered remains of the hut where Murray Levick wrote his Zoological Notes behind me, I am left to contemplate, once more, his strange behavior. When Douglas Russell had found and then published Levick's 1915 manuscript about the sexual habits of the Adelie penguin, it was not just the censorship of his science that caught my attention, it was Levick's own apparent complicity in keeping the information from the public or, indeed, anyone else. Levick's attempt to hide some of the more salacious elements of his original observations by pasting over them using a code of Greek letters spoke of someone embarrassed or unsure about the repercussions of what he had observed and written down. The assumption seemed to be that Levick had done this in retrospect, perhaps when back in England, but I now think not.

All the evidence points to Levick doing this while still in the Antarctic, while still at Cape Adare. The ink in the covered-up pieces is the same as the rest of the notebook's entries and the fountain pen used to make the additions leaves the same telltale markings, especially when writing the letter *k*.

What is certain is that Levick was worried not about the propriety of his data per se, as everything else was left in English but the sensitivity of some of the subject matter. Or perhaps his sensitivity to the subject matter. His Victorian values. His prudish nature.

Despite all that, he reverted to English after his first censored piece, but left little doubt that it was about the sexual behavior of the penguins:

> *The cock did not seize the hen with his beak, by the feathers on the back of her head as chickens do. I was lucky enough to get a photograph of this.*

Tellingly, the word "chickens" had been added in pencil, crossing out was written there already, as if he had gone back to edit it sometime later. There are also four pencil crosses, like large kisses near the bottom of the page, as if he had marked this page for special attention. There is also an added note, written in a different light blue ink that says, "More notes on this later." All this suggests that such editing occurred sometime later, whereas the original cover-up occurred when he still had the fountain pen with the blue-black ink. That is, at Cape Adare itself.

The strongest evidence for this, however, comes from the entry in Levick's Zoological Notes for December 5, 1911. This entry is notable because he wrote in Greek letters but he did so directly onto the page. This was no afterthought following the initial writing, this was a deliberate attempt to obfuscate the content at the time he was recording it, and hence it had to have been done in situ, in the hut at Cape Adare.

At first he wrote in English, "I saw a couple of penguins at an empty nest today, in the midst of a group of occupied nests." Then he switched to using his code of Greek letters: "the hen was sitting on the nest and the cock copulating with her." Then he switched back to English: "I mention this as it is so late in the season and the chicks expected out any day, The fact that the nest was empty being remarkable," which was followed by more Greek letters suggesting, "the copulating is still to be seen occasionally all over the rookery"

The following day, December 6, 1911, he again started to write in his coded Greek letters directly onto the page. But, having written six such words, he crossed them out and continued to write in English instead.

I saw another astonishing sight of depravity today. A hen which had been in some way badly injured in the hindquarters, was crawling painfully along on her belly. I was just wondering whether I ought to kill her or not, when a cock noticed her in passing, and went up to her. After a short inspection he deliberately raped her, she being quite unable to resist him. After this, he had hardly left her before another cock came up, and without the slightest hesitation, tried to mount her. He fell off at first, and desisting, stole two stones from neighbouring nests, and dropped them into a deserted nest just in front of her, after which he mounted and copulated with her. After he had gone, the poor hen struggled about 20 yards, when another cock ran up to her and was just starting in the same way as the other, when a fourth ran up and fought the third cock, driving him away, and afterwards raped her as the others had done. After this, the hen who was much more lively now, struggled on, and had got about ten yards further when no less than three more cocks gave chase, each trying to climb onto her, but this ended in a short fight, after which they went their several ways. The hen lay down, and as the poor thing evidently knew her way, making as she was, in a straight line, I left her, deciding that she might recover if she reached her own nest.

Rape, indeed, gang rape at that, was clearly the nadir, as far as Levick's opinion of the sexual behavior of the penguins went. As he put it so succinctly, "There seems to be no crime too low for these penguins."

Curiously, this was one incident—arguably the most depraved of the sexual behaviors of the Adelie penguins that Levick observed—that he left uncovered, undressed in his Greek letters, although it was clearly his intention to do so when starting the entry. Why change his mind? And, more so: Why not paste over them like he did the other entries he had written in English? Was this because he had so clearly passed moral judgement on it and, therefore, it could not reflect on his own moral stance?

He certainly cannot have been covering up his observations to keep them secret from the other men. In fact, Levick noted that he had deliberately called Campbell over so that he might witness the rape with Levick. Strangely, however, Campbell made absolutely no mention of this in his diary. Perhaps Campbell was too disgusted to record such observations, or perhaps he felt it was Levick's job as the expedition's assigned zoologist? Although, interestingly, Campbell often made zoological-type observations in his diary on other occasions, such as when he had helped Levick kill the leopard seals. I suspect that rape was not a subject he wanted recorded in his diary.

What seems most likely to me as I sit reviewing the evidence in my cabin, is that faced with being picked up from Cape Adare by the *Terra Nova* and transported down the coast to go exploring, Levick's Zoological Notes and photographic plates would need to remain onboard the ship. Perhaps he was most worried about what others might think? In particular, what Wilfred Bruce, the original owner of the fountain pen for which Levick had traded his gun, and the second in command on the *Terra Nova*, might have thought of what Levick had been up to with that pen?

⸎

It is January 7, 2012. Douglas Mawson is writing history: he has discovered a new part of Antarctica with a safe harbor between rocky outcrops and islets, which he names Commonwealth Bay. The head of the bay he names Cape Denison. All around, the rocky shores are covered with Adelie penguins, Weddell seals, and their chicks and pups. It is calm and warm. Mawson chooses to set up his main base here, while the *Aurora* will take a smaller party to set up a second base to the west before returning to Hobart for the winter.

As picturesque and pleasant as Commonwealth Bay is on the day he arrives, Mawson will discover that he has opted to build his hut at the windiest place on earth. That discovery is quick in coming: while they are unloading the ship a vicious wind gives them a welcoming taste of what is to come. The captain of the *Aurora*, John King Davis, will write:

> *Nothing I had experienced in the Ross Sea or in any other part of the world came up to the gales and blizzards of Commonwealth Bay for sudden violence and frequency.*

It is January 8, 1912. After a "pleasant and uneventful trip," according to Priestley, down the Victoria Land coast, the *Terra Nova* reaches Evans Coves in Terra Nova Bay. The bay had been named after the ship by Scott during the Discovery Expedition when the *Terra Nova*, together with the relief ship *Morning*, had come to McMurdo Sound to try to help free the *Discovery*, which was locked in the frozen sea ice at Hut Point. Opposite them now is the 8,990-foot Mount Nansen, which had also been named by Scott during the Discovery Expedition as a way of acknowledging the assistance of Fritjof Nansen. Ironic, really, because at that very moment Scott is struggling to man-haul his way to the South Pole after Nansen's protégé, Amundsen, who using the dogs Nansen had advocated that Scott should use, had already gotten there over three weeks earlier.

When Levick and the others are dropped off, the pack ice is thick and barely passable, at first blocking their way from reaching the shore at all. It takes them several attempts before they are able to find open water and land at Evans Coves. The men transport their gear from the *Terra Nova* half a mile across the sea ice to a piedmont. Then they are left alone with two sledges and enough food and equipment for five of weeks exploring. They are to be picked up by the *Terra Nova* on February 18. They have taken another four weeks of skeleton rations in case the ship should be delayed in getting to them, but this has never been a serious consideration because, as Priestley puts it, "we would all have sworn that if there was one place along the coast which would be accessible in February, this would be the one."

The reason for their unbridled confidence is the Drygalski Ice Tongue lying at the southern edge of Terra Nova Bay. It is the giant glacier of ice that I had witnessed when first flying down to Antarctica; a massive frozen river of ice that pokes out into the Ross Sea for over forty miles from the coast, causing the pack ice and bergs to "bank up" on its southern side, and then to "stream northwards" from its tip, "well away from the land," as Priestly puts it.

Antarctica is not, however, a place where survival is best based upon assumptions.

The *Shokalskiy* moves slowly along the sides of the Drygalski Ice Tongue. This tip of the glacier is enormous, jutting out into the sea for forty-three miles and being up to fifteen miles wide. Up close, it looks like a mini Ross Ice Shelf. On its northern side, we edge around beautifully sculpted bergs: old bergs that have been weathered by the sun and water, reflecting turquoise blue from their undersides. They seem to have been trapped there in the protected eddies on the side of the ice tongue sheltered from the southerly winds that blow from the South Pole. Three Adelie penguins sit on the scalloped blue-green terrace of one particularly astonishing one. The waters of the Ross Sea lap at their feet like it is a Mediterranean resort.

It is a fine day and Terra Nova Bay has a benign feel to it: the exposed rock, the glaciers, the icebergs, and the cone-like perfection of the extinct volcano to the north, Mount Melbourne, make it seem more like a theme park version of Antarctica than, like Commonwealth Bay, one of the most inhospitable places this side of Hell. The open blue waters of the Ross Sea gently slap the rocky shore and the Ross Sea itself has the least amount of pack ice in it since records had begun. Access for us is easy through the open water.

The men of the Northern Party are all excited because here, at last, they can really make their mark. The land north of the Drygalski Ice Tongue is completely unexplored. Campbell decides to split the party in two. He, Priestley, and Dickason shall form the main exploring party, surveying the area around Mount Melbourne and collecting geological specimens. Meanwhile, Levick is to lead a secondary party consisting of himself, Browning, and Abbott to explore the area to the southeast of what would become known as the Campbell Glacier.

It does not all go well. Levick misinterprets Campbell's instructions and fails to meet up with him at their designated rendezvous point. Peeved, Campbell leaves to explore the area around what will become known as the Priestley Glacier. It is there that he makes an important find: a piece of sandstone containing the fossilized trunk of a tree: it is proof positive that in the past Antarctica was covered with forest. The geologist Priestley is ecstatic:

> . . . during the past ages the Antarctic has possessed a climate much more genial than that of England at the present day . . .

Levick's party, meanwhile, blunders along, trying to catch up. In doing so, they get caught up in a wretched field of crevasses. Even so, Levick is happy with the rock and lichen specimens he is able to collect.

And, as these explorers take from the environment, they also add to it. In addition to the Priestley Glacier, the large glacier running southwest from Mount Melbourne is to be named the Campbell Glacier, while three smallish peaks standing in a line to the southwest of Mount Melbourne are named after the three men: Mount Dickason (6,660 feet), Mount Browning (2,500 feet), and Mount Abbott (3,346 feet). Notably, the short Dickason is with Priestley and Campbell and his peak is twice the size of the tall Abbott, who is tenting with Levick. Levick, being an officer, receives an even greater honor: Mount Levick, to the northwest of Mount Melbourne, stands at 7,840 feet.

It is the afternoon of January 16, 1912. The five British men push on. They are now within a day's march of the pole when Bowers spots an unusual mound of ice, which he at first takes to be a cairn, but then dismisses it as strastugi. Thirty minutes later, he sees a black speck and the worst fears of the men are confirmed: it is one of Amundsen's sledge runners with a black flag tied to it. In the snow they see sledge tracks, ski tracks, and dog prints. "This told us the whole story," writes Scott that evening.

At 6:30 P.M. the next day, January 17, 1912, Scott and his men finally arrive at the South Pole, thirty-four days after Amundsen. The complete extent of their devastation and disappointment is captured by Scott in his diary entry that day:

> *Great God! this is an awful place and terrible enough for us to have laboured to it without the reward of priority.*

The next day, their new readings show that the actual pole is located three and a half miles away. They find the Norwegian's tent one-and-a-half miles from their calculated position for the pole, exactly as the Norwegians had calculated it too. Inside Scott finds the note to him and the letter to King Haakon VII, which he takes. He leaves his own note to say they have been there. Then at lunch, when half a mile from their calculated position for the pole, they go about the dreary business of building a cairn, flying the Union Jack, and taking a photo of themselves. If ever a photo is worth a thousand words, Bowers's photo is it: the look of dejection is captured on the faces of five men who seem bereft of elation despite having reached their goal. They look cold, sad, done in.

A half mile later they find another of Amundsen's sledge runners, which Scott remarks:

> *I imagine it was intended to mark the exact spot of the Pole as near as the Norwegians could fix it.*

In an ironical twist, whether from pique or necessity, the Englishmen take the Norwegian's sledge runner marking the exact position of the South Pole, to help fashion a sail for the journey back. They are hoping to be assisted by the nearly continuous southerly winds.

Then they leave, Scott's mood now as low as the temperature.

> *Well, we have turned our back now on the goal of our ambition and must face our 800 miles of solid dragging—and good-bye to most of the daydreams!*

The real problem for Scott and his men is not that they must turn away from the goal of their ambition and their daydreams, it is the eight hundred miles of solid dragging that lie between them and safety. In contrast to the Norwegians, they are weak and suffering because of inadequate rations and a lack of vitamins. They are developing scurvy. The largest of them, Petty Officer Taffy Evans, a mountain of a man, and from the outset the strongest of them all, gets the same rations as the rest of them. If they are inadequate for the smallest of them, they are pitiful for him. He is suffering the most and his condition is exacerbated because Scott's plan had called for him to change the runners of the sledge from twelve-foot ones to ten-foot ones when they were on the plateau; a process that took him many hours to accomplish and damaged his hands badly from the cold. He fingers are covered now in large painful blisters from frostbite.

But there is another problem lurking too: the weather. That window in the Antarctic summer when conditions are relatively benign seems to be closing.

It is now January 24, 1912, and already, in the six days since leaving the pole, they have experienced two gale force blizzards, with the current one keeping them in their tent on the plateau. Bowers writes that they are "thinning" and "get hungrier daily." The worsening weather is now a real worry for Scott:

> *Is the weather breaking up? If so, God help us, with the tremendous summit journey and scant food.*

Scott simply does not have the reserves to allow for such contingencies as delays due to weather.

It is January 25, 1912, 4:00 A.M. Amundsen and his party of four other men, two sledges, and, by then, eleven dogs arrive back at Framheim having traveled very quickly, often covering twenty, thirty, even up to forty miles in a day. They have covered the journey of 1,680 miles in just ninety-nine days. All the men are in good condition. In fact, when Amundsen weighs himself, he finds that he has put on weight during the journey, so well supplied with food have they been. Waiting for them, in addition to the cook, who had stayed at Framheim throughout, are Johansen, Prestrud, and Jørgen Stubberud, who have become the first humans to set foot on King Edward VII Land, having robbed Campbell, Levick, and the others of that honor. Best of all, from Amundsen's perspective, the *Fram* is tied up and waiting for them.

Four days later, on January 30, 1912, Amundsen closes and locks the door of Framheim. They load their thirty-nine remaining dogs on board *Fram* and then cast off, leaving Antarctica to the penguins and setting their compass for Hobart in Australia.

They have spent much of the last few days cleaning Framheim. Unlike the huts at Capes Adare and Royds, they leave theirs spotless. It does not matter. No one will ever visit Framheim again.

It is late 1961. I am just seven, just hearing for the first time—at least the first time that I can remember—a tale of Antarctica presented to me as a grand tragedy in which Scott is the heroic character and a colorless figure called Amundsen plays a bit part as the foreign spoiler.

Sometime around that moment, a large piece of the Ross Ice Shelf in the Bay of Whales, a piece containing Framheim, calves off the shelf. It is similar to the calving of the ice shelf that had seen Borchgrevink's original inlet and Scott's Balloon Bight disappear. The enormous piece of ice breaks into smaller icebergs, which are transported west by the prevailing currents. Eventually, they break down completely.

Framheim now, fittingly, lies somewhere under the Ross Sea, that body of water that had been first discovered by James Clark Ross, who would go in search of John Franklin, whose loss, in turn, would spark the boyhood dreams of Roald Amundsen and eventually bring him here. It was like the sun going full circle around the apex of the pole: a curious sight, indeed.

And the answer to your question, Roald, is no, nothing more topsy-turvy can be imagined.

∽

It is March 2000. The Ross Ice Shelf this time calves off what is the largest iceberg ever recorded. At 183 miles in length, it is about the size of Jamaica. Named B-15 by scientists exercising an objective methodology, of which Levick might have approved, it drifts westward in the same currents that had propelled the one containing Framheim. Eventually, the largest chunk of it, now designated, with equal imagination, B-15A, crashes into Ross Island along with another big berg it has dislodged by clonking into the ice shelf just beforehand. This second, smaller berg goes by the moniker C-16.

Unfortunately for the penguins breeding on Ross Island at Capes Bird, Crozier, and Royds, B-15A and C-16 become grounded; stuck fast. They are so massive that they alter the marine environment, preventing the winds and currents from breaking up the surface of the sea after it has frozen during winter.

As a consequence, the Adelie penguins breeding at the Ross Island colonies must walk up to eighty miles across the ice to get access to the sea and their food. If this is hard for them, it is devastating for their chicks. Parents take much longer on their trips to get to get food: whereas, as Levick had noted, parents would normally be away from the nest only a day or two once the chicks hatch, now they are taking three, four, five, or more days. And when they do arrive back at their nests, they bring much less food in their stomachs that they are able to regurgitate to their starving chicks.

During the 2000–2001 breeding season, there is almost total breeding failure for the penguins on Ross Island. Their penguin chicks are on the receiving end of the same cruel reality experienced by Scott eighty-eight years earlier: Antarctica is no place to be without enough food.

Even though Murray Levick had to leave Cape Adare when the oldest penguin chicks were little more than three weeks of age, he managed to make two remarkably perceptive observations about the change in the penguins' breeding behavior that occurs at that time, which he reported in his book:

> *The first of these is that the chick's downy coats become thick enough to protect them from cold without the warmth of the parent; and the second that as the chicks grow they require an ever-increasing quantity of food, and at the age of about a fortnight this demand becomes too great for one bird to cope with.*

The chicks start to form crèches within the subcolony. "The individual care of the chicks by their parents is abandoned, and in place of this, colonies start to 'pool' their offspring, which are herded together into clumps or 'crèches,' each of which is guarded by a few old birds, the rest being free to go and forage."

However, after this, Levick's scientific logic lets him down. He assumed that the adult penguins cooperated to feed all the chicks in the crèche. Had he marked the chicks and adults, he would have seen that parents feed only their own chicks.

RIGHT: George Murray Levick, Surgeon Commander, in full dress uniform, World War I. *Photo from the British Exploring Society, London, UK.*
BELOW: Lloyd Spencer Davis onboard the *Shokalskiy. Photo from Scott Davis, ScottDavisImages.com.*

ABOVE: Levick's home, Budleigh Salterton. *Photo from Lloyd Spencer Davis.* RIGHT: Murray Levick, self-portrait, on skis near Cape Adare. *Photo from the British Exploring Society, London, UK.* BELOW: Apsley Cherry-Garrard visits the Levicks, 1926. *Left to right:* Audrey Levick, Murray Levick, Rodney Levick, Apsley Cherry-Garrard, Cherry-Garrard's companion. *Photo reproduced with the permission of the East Sussex Record Office, all rights reserved.*

RIGHT: Roald Amundsen in Hobart, Australia, March 1912, after becoming the first person to reach the South Pole. *Photo from The Fram Museum, Oslo, Norway.* BELOW: Amundsen's house, *Uranienborg*, Norway. *Photo from Lloyd Spencer Davis.*

TOP: Emerging from the snow cave, September 1912. *Left to right:* Abbott, Campbell, Dickason. BOTTOM: *Left to right:* Priestley, Levick, Browning. *Photos from Victor Campbell's album, reproduced with the permission of the Queen Elizabeth II Library, Memorial University of Newfoundland, St John's, Canada.*

ABOVE: Base camp of the Public Schools Exploring Society 1937 Expedition to Trout River, Gros Morne National Park, Newfoundland, Canada. *Photo from the British Exploring Society, London, UK.* BELOW LEFT: Cover of Levick's Zoological Notes. *Photo from Lloyd Spencer Davis.* BELOW RIGHT: Pair of Adelie penguins photographed by Levick engaging in mutual calling. *Photo from Victor Campbell's album: reproduced with the permission of the Queen Elizabeth II Library, Memorial University of Newfoundland, St John's, Canada.*

LEFT: Fridtjof Nansen, 1896. *Photo from The Fram Museum, Oslo, Norway.* BELOW: Nansen's desk with picture of his wife, Eva, on the left. *Photo from Lloyd Spencer Davis.*

ABOVE: Nansen's home, *Polhøgda*. BELOW: Nansen's study at the top of the tower in *Polhøgda*. *Both photos from Lloyd Spencer Davis.*

Sketches by Victor Campbell. TOP LEFT: Pair of Adelie penguins on nest of stones. TOP RIGHT: Pair of Adelie penguins with two chicks on nest. CENTER: The *Terra Nova* in Robertson Bay, Cape Adare, with Admiralty Range behind. BOTTOM: The meeting of the *Terra Nova* and *Fram* in the Bay of Whales. *All images reproduced with the permission of the Queen Elizabeth II Library, Memorial University of Newfoundland, St John's, Canada.*

ABOVE: Adelie penguin on ice floe, Cape Royds. BELOW: Borchgrevink's hut and storeroom (without roof), Cape Adare. The remains of the Northern Party's hut are just visible to the left of Borchgrevink's. *Both photos from Lloyd Spencer Davis.*

TOP: Sketch by Victor Campbell of courting Adelie penguins engaged in a mutual display. *Image reproduced with the permission of the Queen Elizabeth II Library, Memorial University of Newfoundland, St John's, Canada.* CENTER: Carsten Borchgrevink, 1897, the first person to stand on Antarctica and the first to winter over on the Antarctic continent. *Photo from The Fram Museum, Oslo, Norway.* BOTTOM: Inside the Cape Adare hut. *Left to right:* Abbott, Campbell, Browning, Levick, Priestley (Dickason obscured). *Photo from Victor Campbell's album, reproduced with the permission of the Queen Elizabeth II Library, Memorial University of Newfoundland, St John's, Canada.*

TOP: Inside Shackleton's hut, Cape Royds. CENTER: The stove and cooking area, Shackleton's hut. BOTTOM: Wall of Shackleton's hut with photos of Queen Alexandra and King Edward VII. *All photos from Lloyd Spencer Davis.*

TOP: Scott's Hut Point hut, McMurdo Sound. CENTER: Boxes from the Terra Nova expedition left in the Hut Point hut. RIGHT: Cooking area, Hut Point hut. *All photos from Lloyd Spencer Davis.*

TOP: Mount Melbourne, Terra Nova Bay. CENTER: Shackleton's hut, Cape Royds, and Mount Erebus. BOTTOM: Scott's hut, Cape Evans. *All photos from Lloyd Spencer Davis.*

TOP: Scott's bed, Cape Evans hut. CENTER: Lloyd Spencer Davis sitting in Levick's chair in his house in Budleigh Salterton, surrounded by his photographs from Cape Adare and his skis. BOTTOM: Adelie penguins mating. *All photos from Lloyd Spencer Davis.*

ABOVE: Adelie penguins on base of iceberg near the Drygalski Ice Tongue. BELOW: Adelie penguins resting on iceberg, Cape Adare. *Both photos from Lloyd Spencer Davis.*

ABOVE: The beach at Budleigh Salterton. LEFT: Douglas Russell with the unpublished 1915 manuscript by Murray Levick that he discovered. *Both photos from Lloyd Spencer Davis.*

TOP LEFT: An Adelie penguin after escaping the jaws of a leopard seal. TOP RIGHT: An Adelie penguin with a stone. CENTER: A male Adelie penguin performing an ecstatic display, Cape Bird. BOTTOM: An emperor penguin walking in front of ice cliffs, Terra Nova Bay. *All photos from Lloyd Spencer Davis.*

TOP: A feeding chase: an Adelie penguin is pursued by its two chicks. CENTER: Joseph Hatch's discarded digesters amidst a colony of King penguins, Macquarie Island. BOTTOM: A Weddell seal, Terra Nova Bay. *All photos from Lloyd Spencer Davis.*

ABOVE: Adelie penguins and chicks, Cape Crozier colony. The icebergs B-15A and C-16 stretch across the horizon and beyond. BELOW: Leopard seal sleeping on an ice floe with the *Shokalskiy* behind. *Both photos from Lloyd Spencer Davis.*

TOP: Leopard seal attacking fledgling Adelie penguin chicks, Cape Adare. CENTER: An adult Adelie penguin lunges at a skua attacking a crèche of chicks. BOTTOM: South Polar skua in attack mode. *All photos from Lloyd Spencer Davis.*

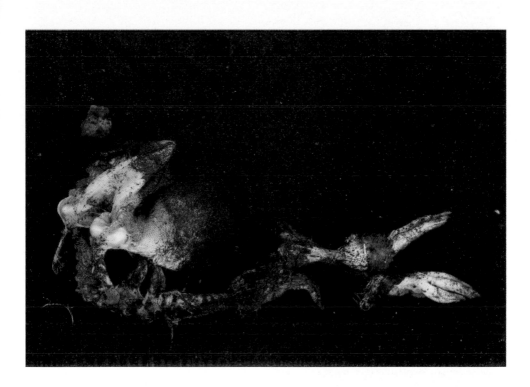

ABOVE: Remains of an Adelie penguin chick after being eaten by skuas. BELOW: A skua dragging an Adelie penguin chick away from the safety of the crèche. *Both photos from Lloyd Spencer Davis.*

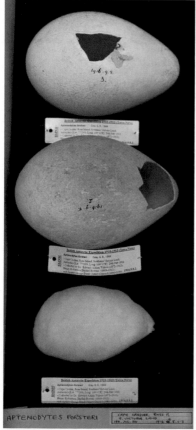

APTENODYTES FORSTERI

ABOVE: An Adelie penguin porpoising. LEFT: Two of the "winter eggs" of Emperor penguins deposited at the British Museum of Natural History by Apsley Cherry-Garrard after "The Worst Journey in the World." OPPOSITE TOP: The Ross Ice Shelf with Cape Crozier and Mount Terror visible behind. OPPOSITE BOTTOM: An Adelie penguin, caught and preserved by Murray Levick and deposited at the British Museum of Natural History. *All photos from Lloyd Spencer Davis.*

ABOVE: Statue of Roald Amundsen on the banks of Bunnefjorden, Norway. RIGHT: Memorial cross for Scott's Polar Party atop Observation Hill overlooking the American base of McMurdo. *Both photos from Lloyd Spencer Davis.*

It is my first summer in Antarctica. 1977. Faced with the necessity of coming up quickly with a research project on Adelie penguins so that I may preserve my chance of going, I choose to focus on the crèching behavior of their chicks. Everything I have done at Cape Bird until the chicks start crèching has really been just filling in time; a prelude to my proposed study.

Crèches are assumed to keep chicks warm and protect them. Not all penguins crèche, but the Antarctic penguins mostly do. In Emperor penguins, it clearly has a thermoregulatory role: even the adults of that species huddle tightly together, which is understandable when breeding in the Antarctic winter.

However, I find that by the time the Adelie penguin chicks are two weeks of age, their finely feathered down insulates them so well that in the relatively benign weather of midsummer in Antarctica, on all but the very coldest days, the chicks hang out as loose coalitions without any bodily contact that could provide thermal benefits. I discover, instead, that the primary determinant of whether chicks will crèche is not the weather, but the number of adults in the subcolony. If there are enough adults around, they may not even bother crèching, no matter how old they are.

The adult birds do not feed any chicks that are not their own, and they may not even chase skuas away that are attacking a nearby chick. However, by their mere presence, the adults provide passive protection when chicks are left in crèches by their parents that are away at sea trying frantically to collect enough food.

What influences the number of adults ashore is a rather curious phenomenon called the Reoccupation Period. This is a time when the number of adults in the colony rises again due to an influx of adult birds that either had not been able to secure a mate during the courtship period or had failed in their breeding attempt soon afterward. These non-breeders and failed breeders come back to the colony about the time that the chicks are hatching and typically stay for several weeks. During this time, they go about the business of "practicing" breeding. They establish nests in the subcolonies. They ecstatic display. They establish partnerships. They even copulate. But they never produce any eggs. It is all a sham; at best, a practice run.

The *Shokalskiy* now is being buffeted by strong winds in Terra Nova Bay, as the Disneyland version of Antarctica is replaced by the real one. Even with a good many of my seasick pills in me, it is all I can do to read Levick's manuscript, the one found by Douglas, one more time.

Levick calls the unemployed birds that frequent the colony during the Reoccupation Period, "hooligans," and as is his way, he assumes that these are mostly "cocks," when, in fact, both males and females show up to the colony during the Reoccupation Period. Levick, however, has nothing but contempt for the behavior of the male unemployed birds, noting that they fornicate with chicks that are without their parents, oftentimes killing them in the process.

In his book, *Antarctic Penguins*, he says that these "hooligans" kill chicks, but as directed by Sidney Harmer, he leaves out all references to their wayward sexual behavior, saying only of the "hooligan cocks":

> *The crimes which they commit are such as to find no place in this book, but it is interesting indeed to note that, when nature intends them to find employment, these birds, like men, degenerate into idleness.*

I am sure that Levick really did witness rape, buggery, pedophilia, necrophilia, and all. But by focusing his attention on their bad behavior, he missed the good that these unemployed birds do from the chicks' perspective. By their mere presence, they keep the voracious skuas at bay.

The bawdy behavior of these "hooligans," as observed by Levick, is likely to be a symptom of the same thing I observed during the real courtship period: males are not very discriminating and will basically bonk anything that moves, and many things that don't move, as well. During the Reoccupation Period, there really are no negative consequences for these adult males if they mate with the wrong thing. Although, if it is really all about practice, there may be a point to trying to mount an actual female.

For male Adelie penguins, jumping onto a prone female's back while vibrating his bill against hers to stimulate her, then padding backward down her back until he reaches the nether region before lowering his tail to bring his cloaca in contact with her cloaca as he shoots his bolt, is not the easiest of tasks. When Fiona and I observed hundreds of penguin copulations as part of our study of the Sperm Wars, we found that about one-third of the males fell off the female before they reached the business end, one-third fired their sperm waywardly like out of control Gatling guns, and only one-third managed to get their sperm on the target or close enough to it for the female's pulsating pink cloaca to suck the sperm into her reproductive tract.

But in the Antarctic, it turns out, there are many ways to get screwed, and the easiest, by far, is by the weather.

CHAPTER SEVENTEEN

WEATHER

It is January 28, 1912. The Polar Plateau. In addition to Evans not being in good condition, Oates's feet are frostbitten and now are really troubling him. Wilson, the doctor, records that Oates's "big toe is turning blue-black." To make matters worse, Wilson himself has snow blindness. The goggles the men have taken are inadequate round little things that cover each eye and fog up easily, encouraging the men to take them off. Amundsen, on the other hand, used an Inuit design modified by his great mentor, Frederick Cook, which were large and covered both eyes like a modern ski goggle, with slits for ventilation so they did not fog up.

Nevertheless, Scott and his men push on; they have no choice. "The weather is always uncomfortably cold and windy," according to Wilson. Yet by the time they reach the Upper Glacier Depot on the Beardmore Glacier on the evening of February 7, the place where the first supporting sledge containing Atkinson and Cherry-Garrard had left them on the outward journey, the temperatures are warmer and the wind has abated. Incredibly, despite Scott noting that Evans "is going steadily downhill," the next day he allows Wilson to take rock samples from Buckley Island and Mount Darwin (really, both are just the tips of great mountains that poke through the top of the Beardmore Glacier), adding thirty-five pounds to the weight of the sledge they are pulling. And this on a day when Evans is unable to pull at all and has to be detached from the sledge.

Six days later they are still making their way down the glacier. They are getting hungry once again, they are on reduced rations and their progress is slower than it needs to be. "We cannot do distance without the ponies," Scott tells his men. Yet still they haul the rocks.

⁂

It is February 11, 1912. On this day, Levick makes a pleasant discovery.

While sledging, he has been reflecting upon his time in Antarctica, in part because of the tedious nature of sledging. "Sledge journeys in themselves are terribly monotonous when there is little change of scene," he writes in his diary, "In fact, when there is no collecting or other scientific work to occupy ones (sic) mind, and one is pulling hour after hour with nothing but glaring white ahead, darkened by snow goggles, it is simply a form of mental starvation." It is while dragging his sledge through crevasse fields in an effort to catch up with Campbell that Levick really starts to appreciate how Antarctica has changed him.

> *I have come to regard my life rather in the light of a very serious play in two acts. The first act was over, and the curtain rung down on it when I left civilisation to enter this blank antarctic. Now comes the interval during which I am given ample time to reflect upon the scenes of the first act—: to perceive its weak spots, and the way in which I as the chief character must avoid making such mistakes in the second.*

While this area in Terra Nova Bay is at last satisfying the Northern Party's mission to undertake new exploration, one consequence of their leaving Cape Adare is that Levick had to bring a sudden halt to his study of the penguins. He is pleasantly surprised this day, therefore, when he and Abbott come across "a beautiful little cove with a pebble beach running down to the sea, and a small Adelie Penguin Rookery." It is, he notes, "somewhat smaller than Cape Royds, and most of the adults seem to have gone, though a good few remain to feed a few crèches of youngsters. A considerable number of adults are still in full moult, and a few have finished moulting." He goes on to observe:

> *The majority of the youngsters look pretty thin, & although some were wildly pursuing old birds and clamouring for food in a very plaintive piping voice, I only once saw one get fed, and I am inclined to think that they are now mostly abandoned by their parents and are reaching the stage in which hunger is driving them one by one into the water in response to the newly found instinct to catch their own food there.*

Perhaps, Levick and the others of the Northern Party should have taken note of the "totally de-downed and nearly de-downed" chicks at the water's edge: they are telling him that it is time to get out of there before the fury of the winter arrives.

As if on cue, the next day it starts snowing and blowing.

⸺

February 17, 1912. "A very terrible day," as Scott observes. Evans is unable to keep up, but Scott is forced to push on to get to the Lower Barrier Depot, the place where the dog teams had turned around on the outward journey.

> *. . . the remainder of us were forced to pull very hard, sweating heavily.*

And still they hauled those bloody rocks.

They stop for lunch, setting up the tent. After they have made and consumed their lunch, Evans still has not appeared and they see him far off in the distance. They ski over to him. He is "on his knees with clothing disarranged, hands uncovered and frostbitten, and a wild look in his eyes." He cannot walk, so Scott, Wilson, and Bowers return to get the sledge, leaving Oates to comfort Evans.

By the time they get him into the tent he is in a coma. He never regains consciousness and dies at 12:30 A.M. In many ways, it is a relief for Scott:

> It is a terrible thing to lose a companion in this way, but calm reflection shows that there could not have been a better ending to the terrible anxieties of the past week . . . what a desperate pass we were in with a sick man on our hands at such a distance from home.

Within half an hour of Evans's death, they pack up and leave, making it first to the Lower Glacier Camp and then, after a few hours' sleep, they push onto Shambles Camp, where they can have their fill of the plentiful pony meat.

However, progress is slow as they go on, skiing and pulling the sledge in deep snow. Scott is worried about the impending change in the weather. "Pray God," he writes in his diary, "we get better travelling as we are not as fit as we were, and the season is advancing apace."

It is Sunday, February 18, 1912. The Northern Party men wait expectantly for the ship to return, but there is no sign of it. Campbell's diary entries have become a litany of ways to report on the blowing wind:

> February 17th: Still blowing hard with drift
> February 18th: blowing so hard the tent poles were bent nearly double

February 19th: blowing as hard as ever again
February 20th, 21st and 22nd: Blowing a heavy gale, some of
the squalls being terrific.

The blizzard lasts for eight days and by February 23, they are starting to face the cruel reality that lies ahead for them. Even at this time of year, when the full force of the winter is still to be felt, they are very cold, and Levick's nose gets "rather badly frostbitten." Levick sums up their dire situation:

> *We are now a little anxious about the ship, which was due on the 18th. If she doesn't come in to relieve us, we shall be a bit hard put to it for the winter, as we have nothing but a few extra sledging provisions with us, and no materials for building a hut, and no fuel, and very few matches.*
>
> *We are discussing various ways and means, and think we shall try and find a drift to burrow into, and kill enough seals for fuel (blubber) & food, to last us till midwinter, when we can sledge along the sea ice to Cape Evans, 200 miles to the South'ard.*

On the morning of the same day, February 23, 1912, the *Terra Nova* has become stuck in newly formed pancake ice as it tries to get to the men in Evans Coves, and is unable to go forward or backward.

Earlier, on February 18 and right on schedule, the *Terra Nova* had tried to get to Terra Nova Bay to pick up the Northern Party, but their way had been blocked by dense pack ice. Then came the blizzard, the same one that had been bashing Campbell and the others as they lay confined to their sleeping bags in tents that were at any moment at risk of being torn to smithereens. The ship was driven north and only when the wind eased was it able to head back toward the Drygalski Ice Tongue.

Now there is a new danger: winter really is coming and the sea is starting to freeze over. With shades of the *Belgica* and the dreadful prospect of spending the winter on board a ship locked in ice, the crew manages to get the ship moving south, though it is not until the evening of the next day that they manage to fully extricate themselves from the ensnaring pack ice. By then, all thoughts of getting to Campbell and his men have been put aside: they beat a hasty retreat to Cape Evans.

It is February 24, 1912, and Scott writes perceptively, "It is a race between the season and hard conditions and our fitness and good food." Two days earlier, after wind-driven snow drift had caused them to miss a cairn and pass one of their old pony camps without seeing it, Scott had ruminated on the ill fortune that the change in weather posed for his own party:

> *There is little doubt we are in for a rotten critical time going home, and the lateness of the season may make it really serious.*

The trouble for the four men left is that the surface is poor and the blizzards of season's end are plainly coming "apace." While the temperatures during the day are bearable, at night when the barrier is in shadow, it is really cold and all the men now have cold feet. Just as plainly, the men's fitness is disappearing as fast as their food supplies and, just as importantly, the fuel they need for cooking and melting ice for water. Each day, the equation changes, with the odds shifting in favor of "the season and hard conditions." Each night, the colder it seems to get: -37°F on February 27, then -40°F the next night, and -41.5°F on March 1.

That day they make it to the Middle Barrier Depot but discover there is a shortage of fuel. They have scarcely enough, Scott realizes, even with the "most rigid economy," to get them to their next depot, the Mount Hooper Depot, which is seventy-one miles away. That is, if the weather conditions remain favorable, which they soon show themselves to be anything but: the temperature again falls below -40°F during the night and the wind in the morning blows strong. However, that is not the worst of it. Oates reveals to the other men the condition of his feet, which so far, he has largely suffered in silence: they have become severely frostbitten in the cold temperatures.

It is February 29, 1912. The Polar Party is not the only one bearing the brunt of the worst of the weather that the Antarctic can pitch at them. Blizzards have kept them "practically confined to our bags for 13 days—a record I believe for any antarctic party, and it has been absolutely miserable," writes Levick. It eases that day, only for them, cruelly, to spy what they think is "smoke on the horizon and under it a small black speck." Instead of their ship, it turns out to be "only an iceberg with a cloud behind it."

The next day, Campbell orders that they start killing seals and penguins with the prospect of a long, dark, cold winter ahead; they have very little rations left. They kill two Weddell seals and eighteen penguins that day. It is a start but neither species are numerous. The wind needs to relent otherwise the Weddell seals will not come ashore—and, for their sakes, it needs to do so very soon.

Campbell, meanwhile, asserting his leadership, countermands all other suggestions for possible shelters and decides that they should make a snow cave by burrowing into a snow drift. He finds a suitable drift on the lee side of a small hill about one-and-a-half miles away on an island that forms the western edge of Evans Coves. They shall come to call it Inexpressible Island on account of its dreadful conditions: relentless high winds and large, rounded granite boulders that make walking over them tricky even without the snow, ice, wind, cold, and darkness. As Levick remarks:

. . . the road to hell might be paved with good intentions, but to us it seemed probable that hell itself would be paved something after the style of Inexpressible Island.

———

I am there on the *Shokalskiy*, trying to get to Evans Coves more than a century after the *Terra Nova* is doing the same. The benign weather we experienced on arrival at Terra Nova Bay had quickly, as if with a magician's wave of his wand, turned to blighted weather. The wind—perhaps the very same one recorded by Campbell—is now "blowing a heavy gale" with "some of the squalls being terrific," even terrifying. Over one hundred miles per hour of terrifying. We stand off Inexpressible Island for a whole day, but that wind is relentless. There is no way we can land.

———

It is February 29, 1912. The *Terra Nova* is once more heading north to try to get to the men stuck at Evan Coves. The ship has been to the Cape Evans Hut to pick up personnel going back to New Zealand, including the dog driver Meares, who had become so upset with Scott's leadership that he had requested he go home early. The *Terra Nova* had also put in to Hut Point where it had picked up Teddy Evans, who has scurvy and is close to death.

The problem facing Pennell is the lateness of the season. The pack ice is closing in all around the ship. The next day the *Terra Nova* tries once more to get to Evans Coves to pick up the Northern Party, but becomes repeatedly stuck in the pack ice. "Campbell's chances of relief are getting woefully small," writes Wilfred Bruce on March 2. That day they manage to make some headway toward the Drygalski Ice Tongue, but again the pack ice eventually blocks their way and shuts down any prospects of getting to the Northern Party. The ship beats a retreat back to Cape Evans.

The *Terra Nova* must get back to New Zealand before its way is completely blocked. Late on March 4, 1912, it steams northward with the forlorn hope of trying one last time to get to the Northern Party. On March 6, they are within eighteen miles of the Drygalski Ice Tongue, but they can get no further: their way is blocked by pack ice, which they only manage to get out of "with much difficulty." They have tried this time "only to make quite certain that we can do nothing, we really know it is futile," writes Bruce.

They give up and Campbell, Levick, Priestley, Abbott, Browning, and Dickason are left at Evan Coves to do what no humans have ever done before: to spend a winter in Antarctica without shelter, food, fuel, or even much clothing. Inexpressible, indeed!

It is March 7, 1912. A sunny day. "Our faces shone in rivalry with the sun," Roald Amundsen, first man to the South Pole, will say after the *Fram* has arrived in Hobart. The banks of Storm Bay are scorched brown from a prolonged drought, but Amundsen's news proves even hotter.

He cables his brother Leon, King Haakon VII, and Nansen. He has been to the South Pole. He has got there first.

Amundsen's message arrives just as Nansen is writing a letter to, of all people, Kathleen Scott. Nansen has never particularly liked Scott, but from their first meeting, he has been attracted to Kathleen, and she, him.

While Robert Falcon Scott has been struggling through the snow—face, fingers, and feet bitten by the ferocious cold—in a race with Roald Amundsen to get to the South Pole, improbably, his wife Kathleen has been enjoying the warm embrace of none other than Amundsen's mentor, Fridtjof Nansen, the tall blond Norwegian hero. In one of those bizarre moments that should make the most shocking of Levick's observations about the sexual behavior of penguins seem chaste by comparison, as Scott is lying in a reindeer sleeping bag, with the sides of his tent snapping loudly in gale force winds and the gnawing pain of too little food coming from his belly, Kathleen is within the warm clean sheets of a hotel bed in Berlin making love to Nansen.

Nansen is writing to Kathleen about their next proposed tryst, this one in Paris. After getting the message from Amundsen, he continues:

> *I think of you and what you may wish, more than of him, and am in a strange mood, unhappy and uneasy. Oh why are there so many difficulties in the world and why is life so complicated?*

Nansen even goes on to say to her, "I wish that Scott had come first." However, it is not enough for Kathleen, who can stomach neither Amundsen's success nor her husband's failure. Whether out of guilt or grief, her affair with Nansen is over.

Indeed, Nansen, life rarely gets so complicated as when you fuck someone else's wife.

Murray Levick never saw the likes of this in what he regarded as his sexually depraved penguins. It is one thing for a female penguin to mate with a male and then switch partners—as Levick and I had both documented. In those cases, there is a kind of handing off of the baton, like a relay race, a sort of sequential monogamy. The type of infidelity displayed by Kathleen Scott and Fritjof Nansen is entirely different: such as when a female goes off for a bit of sex but has no intention of leaving her partner.

Theoretically, scientists argued, it should not occur in monogamous species like Adelie penguins where the males and females look practically identical. Such monomorphism they say, using five syllables when three or less would do, indicates that there is little sexual selection in this species, little reason to think that some types of males do better at the expense of the rest, which would lead to an exaggeration of the traits sought after by the females. Like the way some male deer have massive antlers, or some men are tall and blond and come with a rakish mustache and a Norwegian accent compared to the rather bland, round-faced ones. No, the fact that Adelie penguins look so similar, suggests that virtually all of them must get to breed and will have only one partner at any one time.

Yet, it is 1990s and Fiona and I are in the center of the Cape Bird penguin colony observing the Sperm Wars, focusing on recording copulations, whole copulations, and nothing but copulations. Just as with my earlier work, we record a fair bit of mate-switching during the courtship period, like a penguin version of musical chairs. But we go on to make a remarkable discovery: 10 percent of females cheat on their male partners, whipping away for a quickie with a neighbor then returning to their partners. It is behavior that scientists never even suspected could occur, let alone would occur, in these monomorphic and supposedly monogamous seabirds.

Why would they do that? For the males, no explanation is necessary. We already know they are lacking in any form of discrimination, and from the male's perspective what's there not to like about getting an unexpected copulation? It is a chance to produce offspring yet have another male raise them: a nice bonus, in an evolutionary sense, if you can get it without any cost. But the females are meant to be the discriminating ones: Why risk pissing off your current partner, the one that will rear the offspring, in order to have a bit of sex with another male if there is no intention of switching, no passing of the baton to the new partner?

However, take any group of a dozen male Adelie penguins and it is a fair bet that one of them will be infertile. Hence, from an evolutionary point of view, the advantage to the female penguin of her duplicitous behavior might be to help ensure that at least one of her eggs will be fertilized in the event that her partner happens to be sterile.

That might excuse a penguin, but not a person: Scott had already proven that he was fertile.

Scott and his men soldier on, but Oates is slowing them down, barely able to walk on his left foot, and they are losing precious time waiting for him. Scott hopes that the dogs he asked Atkinson to send will have brought additional supplies to the next depot, the Mount Hooper Depot.

When they get to it on March 9, 1912, it contains less rations than expected. They had been partially used by the other support parties on their return, especially because no allowance had been made for taking the dogs as far as they did. But much, much worse than that: there is very little fuel. The leather seals on the red fuel cans, which had been placed on top of the food depots to aid their visibility from afar, have evidently swelled as they lay exposed to the sun for so many months and the paraffin oil they contained has evaporated. That might have been alright had the dogs come out this far to meet them with extra supplies, but Scott notes fatalistically now, "The dogs which would have been our salvation have evidently failed."

CHAPTER EIGHTEEN

DOGS

It is March 10, 1912, and the dogs are, in fact, not so far away: they are at One Ton Depot, seventy-seven miles to the north. They have been there for six days and now, at 8:00 A.M., Demitri and Cherry-Garrard are getting the two teams of dogs ready to move again. Except that rather than going south in search of Scott, they are heading back to Hut Point.

The sequence of events that have brought them to this point might well form the basis of a comedy were the consequences not so tragic.

To begin with, the last support party to leave Scott, the one with three men, had a devilishly trying time getting back from the Polar Plateau. Eventually, their leader, Teddy Evans, had collapsed with scurvy, close to death. They erected their tent and, while one stayed with Evans, the other, Tom Crean, walked for thirty-five miles to get to Hut Point to raise the alarm.

At that moment, Atkinson was at Hut Point and about to leave with Demitri and the two dog teams to take food and fuel to the Polar Party, just as Scott had requested of him. Instead, they took the dogs to get Evans and brought him back to Hut Point on one of the sledges.

Atkinson, the doctor, needed to stay with Evans if he was to have any chance of surviving at all.

It turned out that there was only one man who could be spared to go with Demitri to meet the Polar Party and that was Cherry-Garrard. This, despite the fact that the nearsighted Cherry is nearly blind without his glasses, and, in his own words, lacked the requisite qualifications:

> *I confess I had my misgivings. I had never driven one dog, let alone a team of them; I knew nothing of navigation; and One Ton was a hundred and thirty miles away, out in the middle of the Barrier and away from landmarks.*

Nevertheless, they had left Hut Point at 2:00 A.M. on February 26 and managed to travel surprising well with their two sledges and dog teams, getting to One Ton Depot on the night of March 3, 1912.

It is what happened next that seems so lamentable, so laughable were it indeed a comedy: nothing.

They had plenty of food and fuel for themselves and for Scott's party to get them from One Ton Depot back to Hut Point. What they didn't have was much dog food. With all Scott's confused changes of orders, it turned out that Meares had not depoted any dog food at One Ton Depot, as they had expected. The situation was also exacerbated the day after they got to the depot because Demitri came to the young and inexperienced Cherry-Garrard, who was in charge, and asked that they increase the rations for the dogs because they were "losing their coats." This left them with only thirteen days of dog rations and Cherry-Garrard wanted to allow eight for getting back to Hut Point. In the six days they were hanging out at One Ton Depot, at least two were fine for traveling further south, even by Cherry-Garrard's own admission, and the others were probably doable for experienced dog drivers like a Demitri or a Meares. Furthermore, Atkinson had told Cherry-Garrard before he left Hut Point that if Scott and his men were not yet at One Ton Depot when he got there, Cherry-Garrard himself was to decide what to do.

If he went south, Cherry-Garrard rationalized, because of his poor navigational abilities, he might have missed the party on the two good days; and, because of the wind-driven snow on the other four days, "the chance of seeing another party at any distance was nil." He had calculated that the available dog rations would allow him to go for only a day's march anyway, unless he killed some dogs to feed to the others; something he was loath to do, especially as Scott had said he wanted the dogs in good shape for the following spring sledging trips that were planned.

And so Demitri and Cherry-Garrard simply sat at One Ton Depot, waiting for Scott. Until now, March 10, when they are turning for home. Ironically, they make "23 to 24 miles (statute) for the day." It would have been more than enough had they taken some food and fuel that far south.

What seems so unfathomable is that the men with Scott include the two men with whom Cherry-Garrard had taken the perilous winter journey; the two men to whom he certainly owes his own life. Together, they had crossed the line that separates life from death, yet somehow returned. Surely, he owes those men whatever risks it would take to go south on the basis that, perhaps, they are again near that line.

But Cherry-Garrard has managed to convince himself that his friends and his leader are doing fine.

———

It is March 10, 1912, and the awful stark reality for Scott's party is that if they move only at the pace Oates can manage, they are all dead men. As Cherry-Garrard and Demitri head back to Hut Point, Scott assesses clinically the risk that Oates poses for the rest of them:

> . . . if he went under now, I doubt whether we could get through. With great care we might have a dog's chance, but no more. The weather conditions are awful, and our gear gets steadily more icy and difficult to manage. At the same time of course poor Titus is the greatest handicap.

The following morning, March 11, 1912, Scott orders Wilson to "hand over the means of ending our troubles." Reluctantly, the very religious Wilson gives each of them thirty opium tablets, leaving himself with a tube of morphine.

—∞—

It is March 12, 1912. The *Aurora* arrives back in Hobart, having dropped Douglas Mawson's main party at Commonwealth Bay and his Western Party on an ice shelf, which they have named the Shackleton Ice Shelf. As the *Aurora* motors past the *Fram*, which is still in the harbor, John Davis and his crew come out on deck and give three cheers for Amundsen.

Later, Amundsen gifts twenty-one of his Greenland dogs to Mawson, including one that he had taken to the South Pole and back, which Davis will take down to Mawson on the *Aurora*.

The dogs are coming for Mawson, at least.

—∞—

By now, the scientist has become the predator. With hope of the *Terra Nova* coming all but gone, Levick and the others set about killing as many seals and penguins as they can get. For those penguins remaining at the little colony Levick had discovered, it is not the impending winter that they need worry about. By March 14, 1912, the men have killed 118 Adelie penguins—all those they can get their hands on—and added them to a food larder. They have also killed nine Weddell seals, of which they have eaten two. Even so, the prospects of getting enough food do not look good: Levick estimates that they shall need twenty seals, and as long as the wind continues to blow relentlessly, the seals choose not to come ashore.

Meanwhile, the men also work at digging out their snow cave using ice axes and whatever else they can use to fashion its sides. All the while, the wind blows. On March 17, 1912, Campbell, Priestley, and Dickason move into the still-unfinished cave. Meanwhile, Levick, Browning, and Abbott remain camped at the place they call Hell's Gate—the place where they had first landed at Evans Coves—in order to watch for the ship and kill and cut up the seals and the penguins. These men had become the butchers.

For the first time in ten days, Levick writes in his diary:

> *Abbott, Browning & I have killed & butchered 8 seals, and we have killed about 100 penguins: all the moulting birds that remain. I think we ought to have 20 seals to last us till the spring.*

Levick does not look forward to what lies ahead of them: "it is going to be a queer time for us through the dark months." And, given the conditions right now, God knows what it will be like then.

> *The wind has blown now from the S.E. for a whole month excepting one day, and we are all exasperated, it is so damned cold and miserable, and frequently prevents our getting about at all when we ought to be getting on with the sealing and work on the cave, and we are losing the sun daily.*

As Cherry-Garrard, Wilson, and Bowers found out the hard way, Emperor penguins are able to live and breed during the Antarctic winter. However, they are only able to do so because of some unique physical and behavioral adaptations. To begin with, they are big—huge, by penguin standards, with one Emperor weighing more than seven Adelies stacked together—reducing their surface area to volume ratio and thereby helping to minimize heat loss. When it gets really cold, they huddle together, shifting around so that the individuals on the outside get to take their turn on the inside where they benefit from the huddle's collective warmth and protection from the wind. They can only do this because they do not have a nest site. Instead, they carry their eggs on their feet. This means having a clutch of only one egg, and a big egg at that. They breed on the sea ice in the lee of ice cliffs, but it must be sea ice that will remain secure throughout the winter. Food is likely to be far away, requiring them to walk many miles, perhaps ninety or more, across the frozen sea ice to open water. They manage this constraint only because they are so darn big, with the males being able to go for three months in the cold Antarctic winter without food in order to get through the courtship period, as well as incubating the eggs on their feet for the entire two months it takes to incubate the eggs. This gives the females time to walk to the open sea, fatten up, and then walk back to the colony with food for the newly hatched chicks. The males even have an incredible insurance policy should their partners be a bit late getting back with food for their chicks: incredibly, the males can break down their own body tissues to manufacture a form of food. Daddy's milk.

Even if humans like Levick or the small Adelie penguins could somehow withstand the Antarctic winter's cold, there is no way that either could go without food for long. For the Adelies, this means that when winter arrives they must get away from the icebound Antarctic continent and go to areas where there is open water and access to the food they need.

But where might that be? Ever since Scott's Discovery Expedition, we had known that Adelie penguins inhabit the continent only during the summer months, but we had no idea how far they went or where. In the dark Antarctic winter, in an environment sheathed in ice and icebergs and subject to vicious storms, following the penguins would seem to be an impossible task. Except, potentially, it could be accomplished by satellites. The big breakthrough came in the 1990s with the miniaturization of electronics and improvements in battery technology needed to power a transmitter that could send a signal to satellites passing overhead more than five hundred miles above the Earth.

It is 1991. I am at Cape Bird as late in the season as it is possible to be. Each summer season, there comes a day in mid-February when the American and New Zealand Antarctic programs must shut down their helicopter operations. I am scheduled to be pulled out of Cape Bird on the last day scheduled for flights.

It is a weird place at that time of year. Instead of bustling activity and noise as during the heart of the penguins' breeding season, it is quiet and almost deserted. The subcolonies have all lost their structure, the stones that lined their nests now scattered thither and yon. There is a smattering of penguin chicks left. Most are "de-downed," as Levick would have described them, having grown their survival suits of feathers that will enable them to endure being immersed in the cold Antarctic waters for long periods. Their backs are blue-black, the same shade as the ink in Levick's Zoological Notes. Their chins are white. Some still sport a few tufts of down on the tops of their heads, but otherwise they are slim versions of an adult penguin, ready to prove themselves in the watery world beyond.

Thick snowflakes have started to fall from skies black with winter's clouds. The odd down-covered chicks, pathetic runts, stand immobile, their backs plastered with snowflakes, as they wait for death.

The beach has changed too. Gone is the barrier of push ice, that ten-foot-high jumble of huge ice blocks thrown up by winter's storms, which normally separates sea from land. The exposed beach is black shingle, although there are clusters of small rounded pieces of ice bobbing in the waves and being tossed onto the beach like pieces of flotsam.

The adult penguins that remain at Cape Bird, like the chicks, are largely immobile too: they are molting. Mostly they hang out in little valleys at the base of the cliffs, their loosened feathers carpeting the snow around where they stand, unmoving, conserving all the energy they can. Producing a new set of feathers is energetically very expensive and, while they shed their old feathers and wait for the new lot to grow, their insulation is much reduced. Bereft of effective insulation, the molting penguins are unable to go to sea to feed. Instead, they try to stand close to the cliffs, out of the wind.

Most of the sixty thousand adult birds that breed at Cape Bird do not, in fact, molt there: they leave the colony at the end of the breeding season, fatten themselves up at sea, and then molt elsewhere. That might be on pack ice, icebergs, or ashore at other colonies. When Borchgrevink became the first person to stand on Franklin Island, he discovered massive numbers of molting Adelie penguins there, suggesting that it is a popular molting location for at least some birds in the Ross Sea area. Nevertheless, at Cape Bird, a few adults remain to molt in the colony and, it is possible, some may have come from elsewhere.

I find two adults that have completed their molt and, with help from my colleagues, I attach transmitters to the new feathers on their lower backs using a combination of an epoxy resin and a special tape. It is snowing lightly and it is cold work. I press the transmitter in place with my bare fingers: the five-minute epoxy glue takes more like fifteen minutes to cure in the freezing conditions.

Each transmitter has cost nearly as much as my car. It is unlikely we will ever retrieve them. Nobody has ever done this before. It is unknown if the glue and tape can hold the transmitters to the feathers for the whole winter, and even if they do, the batteries are not strong enough for them to function that long, hence, we will have no way of finding the birds again, except visually. Even assuming these birds return to Cape Bird, with over sixty thousand birds in the colony, finding them will be like looking for a needle in a haystack.

No. As I walk away from the birds, which are now standing unfazed with a streamlined bump on their lower backs, I know that I am consigning the equivalent of two cars to spend their future in the same graveyard that holds Framheim: the bottom of the Ross Sea. Such can be the price of science.

That done, we pack our bags just in time as the helicopter lifts us out of there with winter's telltale storms in cold pursuit.

⁂

It is March 17, 1912. Scott's men are camped, forestalled by a blizzard that is raging outside. It is very cold; now, even at midday, the temperature is -40°F. It is Oates's thirty-second birthday. He had gone to sleep, perhaps with some assistance from Wilson in the form of morphine, hoping not to wake up. But wake up he does. Somehow, he manages to extricate himself from his sleeping bag and crawl across the legs of his companions. Somehow, with his badly frostbitten fingers, he manages to untie the cords that keep the tunnel-like door to the tent shut tight. According to Scott, "We knew that poor Oates was walking to his death" and "tried to dissuade him." Perhaps not emphatically, and certainly not effectively. Somehow, Oates manages to crawl through the tunnel door in the sleeping socks that cover his blackened gangrenous feet. He has no need of boots. He has no need of feet. He is not coming back.

Scott records Oates's last words in his diary:

I am just going outside and may be some time.

That is the last they ever see of him. His body has presumably been covered by the snow from the blizzard when they are finally able to pass through the tunnel themselves and resume their march. They jettison Oates's sleeping bags, a theodolite, and camera, but, remarkably, "at Wilson's special request," Scott consents to the remaining three of them continuing to pull the thirty-five pounds of geological specimens.

<hr />

Nothing illustrates the differences between Amundsen and Scott more than this: for one, the fact that dogs are the proven means of polar travel and, for the other, the notion that there is something more noble, more grand, in man-hauling. Amundsen had traveled to the South Pole and back with speed and little risk either from a lack of food or meeting the fast approach of winter. He used dogs and skis and never man-hauled a single thing, not even a single mile. And here Scott is, being belted by the first storms of an approaching winter, running out of fuel and food as fast as he is running out of fit companions, and yet he finds noble purpose in still pulling some rocks.

There is something about these dogs from Greenland and Siberia that connect the world's two polar regions in a way that even Ross and

Crozier and Amundsen cannot. It is all about evolution and efficiency. The dogs' coats, feet, and temperaments have, through a process of evolution enhanced by breeding, made them able to withstand cold, travel in snow, sleep under snow, and eat the wildlife. Their pack-like ancestry made them trainable, malleable, and able to work together as a group given the right dominance. Of course, that all began in the Arctic, where the Inuit had come to rely on them. In the Antarctic, of course, the dogs are not native but they might just as well have been: all the same things applied, right down to being able to be fed on the local wildlife—in this case, seals—unlike ponies and mules, which needed to have all their forage shipped in.

For that reason, the use of dogs persisted in the Antarctic as the most efficient, most evolved means of transport—like a penguin's streamlined body for swimming in the sea—for more than eight decades after Borchgrevink landed the first dogs at Cape Adare in 1899. In fact, they would probably still be there were it not that one of their advantages—that they eat the local wildlife—came to be seen as a negative.

It is January 28, 1985. The New Zealand Antarctic Research Program is one of the last of all the Antarctic programs to use dogs, and they are to stop within twelve months. Each year, over fifty Weddell seals must be killed to supplement the diets of the dogs, and the public outcry has, understandably, turned public opinion against using dogs in Antarctica, no matter how efficient they are for travel, even compared with more modern means like snowmobiles and caterpillar tractors. There is also concern that they are non-native species, and worse, that they might bring diseases with them to Antarctica.

The first day I arrived in Antarctica, October 18, 1977, a dog team pulling a sledge was there to greet me and a fellow kiwi passenger and take us to New Zealand's Scott Base, while the Americans were all loaded into a big green military vehicle, with wheels as big as I was tall, to go to the American base of McMurdo, a couple of miles over the hill from Scott Base.

I remember how romantic it felt, being with the dogs. You were so right, Cherry-Garrard, this really is adventure, my boyhood dreams galore. It was still early enough in the season that the sun set behind the mountains, and that evening I wandered along the line of dogs chained by the pressure ridges, each bathed in a golden light that highlighted their thick black, brown, gray-and-white fur. They were big animals, boisterous, and with big brown eyes. It was impossible not to love them, to think of them as more pets than beasts of burden. But then you notice that the spacing of their chains is deliberate and far enough so that they cannot tear each other apart. You are reminded that these are animals that will cannibalize their own for survival; an asset if you are Amundsen, a source of revulsion if you are Scott, and in that, he was not alone.

But now it is January 28, 1985, my last day in Antarctica this season, and it will be the last time I ever see the dogs. My team and I must be taken to our waiting Hercules aircraft, which has landed using skis on the permanent ice runway sited on the barrier, the giant shelf of ice over which Amundsen and Scott had both traveled on their way to the pole. The dog handler offers to take us out to the plane using the dogs.

I sit near the front of the sledge. Once again, the sun is low and the dogs are bathed in a soft yellow light. There are eight dogs pulling the sledge and just as a dog chasing a ball can exude enjoyment, the enthusiasm of these animals as they trot along over the snow-covered surface, tails aloft, speaks of joy. It is what they have been bred to do.

There is no shouting from the sledge master, no cracking of a whip. What I remember most is the serene silence of it all. The huffing of the dogs, the creaking of the runners, and the squeaking of the snow: they all made sounds, yes, but they did so in a way that accentuated the silence of being out there on the barrier, moving smoothly across the ice and snow.

It remains the most beautiful, most in touch with nature form of travel I have ever used. And, yes, I love Weddell seals and understand the arguments against using dogs in Antarctica in our modern era, but I would be lying if I said I was not sorry to see the dogs go.

The very next year at Cape Bird, I have the opportunity to do some man-hauling. A crabeater seal, usually a creature of the pack ice, has come ashore and died. A colleague of mine wants the carcass. The only problem is the dead seal is over two miles from the helicopter pad. It is still early in the season at Cape Bird and the beach is still covered in ice and snow. One of my assistants and I fashion a crude sledge by converting my generator-box-cum-hide, and we opt to man-haul the seal to the helo pad. An adult crabeater seal can weigh up to 660 pounds, although they are more typically just under 500 pounds. This one has been there for a while and is desiccated, presumably losing some portion of its weight. Even so, to us it feels every ounce of 660 pounds. Man-hauling is no fun. There is no romance, nothing noble about it that I experience. If I could have thrown away thirty-five pounds of it to ease the load, I would have. If I were Scott, and my life was at stake and I was hauling the extra weight over hundreds of miles rather than just two, I would have done it in a heartbeat.

<hr />

It is March 18, 1912, and the conditions for Levick, Abbott, and Browning camped at Hell's Gate go from bad to worse. "The wind increased to hurricane force, and suddenly one of the tent poles (on the lee side) broke with a snap, and then two others followed, and in a moment the tent was down on top of us, and we pinned down into our bags by it, with a fearful weight of wind on it."

Things do not look good for them. All their belongings are loosely scattered throughout the tent, they are not in their wind-proofs, and the pressing of the tent on them "produced a helpless suffocating sensation." Abbott and Levick struggle into their wind-proofs and crawl out from under the tent, leaving Browning lying on their sleeping bags to stop them blowing away. The wind is so strong that they have to crawl on all fours but they are unable to find anywhere sheltered enough to give them "the ghost of a chance" of erecting their spare tent. With severely frostbitten faces, they crawl back under their tent, and there they wait for the worst part of seven hours for the wind to abate, which it does not.

Finally, they crawl out from under their collapsed tent, pile stones on top of it to prevent their sleeping bags from being blown away, and then make their way on hands and knees toward the snow cave. It takes them one and a half hours to cover the half mile, during which they often need to lie flat on the polished ice surface to prevent themselves from being blown backward. "I shall always remember the appearance of Brownings (sic) face," wrote Levick afterward, "which was dusky blue, streaked with white patches of frostbite, and I suppose the rest of us were the same."

The others revive them with a hoosh and, as they do not have their sleeping bags with them, the men are forced to sleep two to a bag. Levick, who is chided mercilessly by the other members of the Terra Nova Expedition for being fat, gets to spend "a most uncomfortable night" according to Campbell, sleeping in Campbell's bag, who the next day declares, "I was squashed flat."

That day, March 19, the wind moderates enough to allow Abbott, Browning, and Levick to fetch their sleeping bags. From then onward, they inhabit the snow cave, or "igloo," as they call it, for what will inevitably be the longest, darkest, and coldest of winters.

Levick is not looking forward to it:

> . . . the prospect of the winter before us is enough to give anyone the hump I should think . . .

⸺

It is the night of March 19, 1912. Scott, Wilson, and Bowers have gotten to within eleven miles of One Ton Depot. They "have two days' food but barely a day's fuel." It is -40°F and a blizzard blows, keeping them in their tent. Scott's right foot is badly frostbitten: "Amputation is the least I can hope for now." Wilson and Bowers plan to leave Scott alone and go to the depot to get fuel.

It is here that the reality of the short Antarctic season really bites. For ten days, the blizzard rages unabated and "outside the door of the tent it remains a scene of whirling drift."

It is March 29, 1912. It is the end.

CHAPTER NINETEEN

WINTER

The men continue to work on the igloo after shifting into it. The main room is nine feet by twelve feet, with the ceiling being about five and a half feet at its highest point. Naval disciplines are maintained. With his heel, Campbell, the commanding officer, draws a line down the center of the snow cave: one side to mark the mess deck for the men, one side to mark the quarter deck for the officers. The sleeping bags of the three officers are arranged on the right side of the room: first Priestley, who has their precious few stores stacked at the end of his bed to make a small alcove, Levick in the middle, and Campbell at the far end. On the left side there is the small cooking area with its blubber stove, then, somewhat more cramped, the bags of Dickason, Abbott, and Browning. Campbell decrees that what is said by the men on their side of the line is completely up to them and cannot be used against them. Similarly, what the officers discuss on their side of the line cannot be used by the men against the officers. Of course, in the cramped space, they can all hear the conversations of everyone. At first, Campbell insists that they get out of their bags for meal times, but fairly quickly he relents due to the cold, allowing everyone but the cooks to eat in bed. He also insists on having church on Sundays, when he reads from the New Testament and the men sing hymns.

The passageway includes two doorways, constructed using packing cases with canvas flaps, through which the men must crawl. There is a stores area and steps up through a tunnel leading to the outside. They line the internal walls of the main room with blocks of snow that go only half way up the walls, thereby creating a shelf for their meager belongings.

They lie squeezed together on the floor of the snow cave, which they have covered in pebbles and seaweed. Each day, a pair of them makes the morning and evening meals using seal blubber for fuel, which burns slowly emitting a thick black smoke that coats the insides of the cave, the men, their clothes, and everything they touch. At night it is pitch black, but mostly during the days, they lie in what Priestley describes as the "visible darkness" afforded by the faint light emitted from blubber lamps.

Levick has devised the blubber lamps by suspending a string from a safety pin into a tin of oil derived from heating seal blubber. They have a Primus stove with them but very little Primus fuel, so they have crafted a similar arrangement to make a blubber stove.

Levick, aware of the dangers of ill-prepared food, institutes strict rules about the preparation and cooking of their mainly seal meat diet. As it turns out, seal meat is high in vitamin C, which should save the party from dying of scurvy, as Owen Beattie determined had been the fate of the Franklin Expedition. But their acidic diet, with its lack of carbohydrates, has its downsides too: the men are unable to control their bladders. As a consequence, they often wet themselves and are forced to lie in their wet clothes and sleeping bags. Even after placing a can beside each man's bag, they often cannot make it in time; although they tend to keep such accidents to themselves and lie there in discomfort and silence. All of them, at times, experience diarrhea too, with Browning being plagued by severe diarrhea, sometimes needing to "turn out" eight or nine times in a day. Campbell went outside the cave on one occasion to relieve himself but received such severe frostbite on his genitals that thereafter they fashioned a small alcove near the entrance to the cave and have stuck to relieving themselves there.

Levick describes the misery of their living quarters:

> *We are settling into our igloo now, and a dismal hole it is too. Our diet of seal meat is producing curious symptoms. Owing I suppose, to the highly acid state of our urine, we have not only the greatest difficulty in holding it, but some of us are actually wetting our sleeping bags during the night in our sleep. Campbell especially suffers from this, and also from bad haemerrhoids.*

It is almost impossible to imagine their discomfort and squalor: afflicted by acute diarrhea, wetting their clothes and bedding, and no possibility of cleaning themselves, their bedding, or their clothing, and no possibility of changing them either. Their only option is to lie in the filth, cheek by jowl, their clothes and bedding damp with feces and urine, and in Campbell's case, blood too. "Dismal hole" does not seem to even begin to describe it.

It is January 12, 2005. I am in Antarctica with my son, Daniel. Before we can head out to Cape Bird, we must undergo survival training, which is carried out in a crevassed area not too far from New Zealand's Scott Base. We get roped up. We have high-tech clothing, specially insulated boots, crampons, harnesses, carabiners, synthetic ropes, and jumars—mechanical devices for ascending a rope. We know we are going to be gently dropped into a crevasse and then need to climb our way out. Even so, with all our modern gear and our readiness, it is neither psychologically comfortable nor physically easy. I shudder to think how the likes of Scott's men coped with the frequent unexpected falling into crevasses with the crudest of gear.

To make a shelter, Daniel and I construct our own snow cave, piling up shovelfuls of snow to make a drift, then hollowing it out using the shovel. In the twenty-four-hour daylight of an Antarctic summer, it is a surprisingly comfortable residence for a single night: the light shines translucently through the walls of the snow cave. We have insulated mattresses and double sleeping bags packed with pure down from eider ducks. The air temperature is cold, but not too cold. We are snug. The most surprising thing of all: it is so quiet, insulated from any outside sounds, like the wind. That was the first thing that Campbell noted about their igloo.

But make no mistake: swap it for twenty-four hours of darkness and much colder conditions, make it for seven months and not one night, take away the freeze-dried beef stroganoff and frozen prawns, and the novelty would soon go away. It would quickly become a dismal hole.

And having been afflicted by diarrhea when camped in a polar tent, unchanged in design from those used by Scott, let me just say that there are few activities more unpleasant than sitting over a bucket with your trousers pulled down when it is -20°F and blowing a gale.

As with Antarctica's penguins, fat would be an ideal asset for the Northern Party to survive in Antarctica. However, Levick and the others are beginning their vigil in poor condition, exhausted from five weeks of hauling sledges, and they have precious little food. As the doctor in the party, Levick takes it upon himself to ration the food strictly and to keep track of everyone's health. The regimented nature of the party under Campbell's leadership probably helps with the discipline of eking out the merest of rations.

The real issue confronting Levick and Campbell, however, is that it is simply not enough to get through the winter months of darkness. Somehow, they have to hold back enough food, enough equipment, and enough suitable clothing in order to undertake the two-hundred-mile journey back to Cape Evans when, and if, they eventually emerge from their winter hibernaculum. To do otherwise would be to merely delay their inevitable deaths. Even so, Levick has to ensure that they get through the winter in a condition where they can still walk, let alone pull sledges for two hundred miles in some of the toughest conditions imaginable. It is their only assured means of getting to safety. They cannot just sit there and hope to be rescued before their food runs out.

Hence, from the first day they contemplate a winter in Antarctica with little rations, Priestley is put in charge of their meager food supplies with a clear instruction that he must ensure they have enough in reserve to make the journey south when the sun returns in the spring. In fact, while Priestley acts as quartermaster and doles out the food, it is Levick who fashions their diet and decides how much and of what they shall eat.

Campbell's insistence upon stowing the best of their clothes and the best of their food in order to allow for spring sledging seems like an extraordinary piece of foresight when their feet and fingers are freezing and their bellies are empty. It is not unlike the equations the penguins must solve for their winter migrations. They must make sure they have new clothes (feathers) and adequate reserves, not just to exist but to propel them through their long journeys.

⸎

It is the winter of 1991. The French Argos satellites, orbiting the Earth every one hundred minutes, are picking up signals from the transmitters glued to the backs of the two penguins from Cape Bird. They are able to record the position of the penguins to within a mile or two of accuracy. The birds are following roughly the same path that the *Terra Nova* took when getting out of Ross Island for the winter. They travel up the Victoria Land coast, passing the Drygalski Ice Tongue, passing Evans Coves, passing Cape Hallett, moving north at an extraordinary average of over thirty miles each day. When they get to Cape Adare at the outer tip of the indentation that forms the Ross Sea, they hang a left and move westward along the part of Antarctica that had proven off limits and unexplorable to Borchgrevink, Campbell's men, and Mawson. At a certain point, when opposite another massive ice tongue similar to the Drygalski, they turn north, appearing to slow down then and spend the winter fishing in the pack-ice northwest of the Balleny Islands.

All signals are lost from the penguins after five and a half months, most likely because the batteries have become exhausted in the frigid waters, but the data they have furnished are staggering. The penguin that was tracked for the longest traveled over nine hundred miles from Cape Bird to its winter feeding grounds: farther than Amundsen and Scott's journeys to the South Pole. Except that it did not move in a straight line. The actual distance traveled by the bird was 1,735 miles. Given that the birds must make the return journey to Cape Bird for the start of the new breeding season, that means that their winter migration is likely to mean traveling over three thousand miles. Not bad for a knee-high bird that cannot fly.

I think back to that first penguin I met on that October evening at Cape Bird in 1977. If I had known what I know now, I would not have just bowed, I would have gotten down on my knees to worship this tiny creature that had just returned from navigating its way over three thousand miles through the dark, ice-filled waters of Antarctica in winter. Incredible. No wonder so many do not make it.

It is May 1, 1912. Cherry-Garrard, Demitri, and Atkinson are traveling with the two dog teams from Hut Point to Cape Evans.

Earlier, on April 17, Atkinson had taken three other men with him and left Hut Point on a brave, probably foolhardy, attempt to get to the Northern Party by traveling up the coast like the migrating penguins. The sun had all but disappeared for the winter, it was dark and cold, and the ice conditions were treacherous. After three days' hard slog, they got only as far as Butter Point on the other side of McMurdo Sound, where they found the sea ice ahead of them was breaking up. They could not go north to meet Campbell's men any more than the Northern Party could get south to meet them. They depoted supplies at Butter Point and returned to Hut Point. Campbell's party would have to make do by itself.

It is hard going in the gloom on an ice surface that makes it difficult for the dogs to pull the sledges. One dog, Manuki Noogis, lies down and refuses to go on; they cut him loose hoping that he will find his own way back to the hut at Cape Evans, but he is never seen again. As they near Cape Evans, Atkinson asks Cherry-Garrard if he would go for Campbell or the Polar Party come the spring. Cherry-Garrard does not hesitate, "Campbell," he replies.

> . . . just then it seemed to me unthinkable that we should leave
> live men to search for those who were dead.

Yet, the responsibility for finding out what has happened to the Polar Party and whether they reached the pole weighs heavily upon Atkinson. On June 14, 1912, a week before Midwinter's Day, he calls a meeting of all at Cape Evans to discuss their options come the spring.

The thirteen men are all that remain at Cape Evans from the Terra Nova Expedition. The hut is hushed. Any frivolities replaced now by the seriousness of the decision confronting them. Atkinson addresses the men. There are simply not enough of them to mount two search parties. Either they go south to try to ascertain what became of Scott and the Polar Party or they go north to try to rescue Campbell and his men, who, potentially at least, might have survived the winter on seals and penguins, however small that chance might be; unlike the Polar Party members, who have no such chance. On the one hand, they might go south and fail to find any trace of the Polar Party, while Campbell's men "might die for want of help." Cherry-Garrard reiterates the moral dilemma they face, much as he had expressed it to Atkinson on the journey from Hut Point to Cape Evans six weeks earlier: "Were we to forsake men who might be alive to look for those whom we knew were dead?"

Atkinson points out that even if they could get to Campbell, it is likely they could do so only five to six weeks before the *Terra Nova* should return and be able to pick them up anyway. He expresses his own conviction that they should go south. The primary purpose of the expedition has been to get to the South Pole. He suggests that they owe it to the men of the Polar Party and their relatives, as well as the expedition as a whole, to ascertain their fate if at all possible. He puts the two options to the vote. All, including Cherry-Garrard, vote with Atkinson. Only Lashly abstains and refuses to vote at all.

Like Manuki Noogis, then, the decision has been made to cut Campbell's Northern Party loose and leave them to find their own way back to the Cape Evans hut. The dark truth that settles uneasily on them more certainly than the long black night outside, is that none of them know if, like their dog, they will ever see Campbell, Levick, Priestley, Abbott, Browning, and Dickason again.

The night is at its longest and blackest. It is June 22, 1912, Midwinter's Day. In the snow cave at Evans Coves, Levick describes it as "a great day of feasting." He and Priestley are the cooks. They prepare "the most magnificent hoosh, with emperor penguins (sic) hearts and livers, and seals (sic) liver, meat, and plenty of blubber." This is followed by a "brew of Fry's Malted Cocoa of the actual strength employed in civilisation, and I think we enjoyed this more than anything." After this comes "four biscuits and four sticks of chocolate each, and 12 lumps of sugar."

I have not realised how hungry I have been during the last month or so, till I felt the relief of a full stomach and the absence of a worrying appetite.

They also drink their only alcohol, a bottle of British fortified wine called Wincarnis. Priestley says, "none of the famous wines of the world could possibly taste to us as did this," although, while chopping meat, he knocks over his mug of the precious wine, spilling it over his sleeping bag. He manages to save less than a tablespoonful, although he good-humoredly maintains this enhances his appreciation of the little wine that remains. Yet like Levick, it is the full-strength cocoa that is his pick of the meal:

> *The hoosh flavoured with seal's brain and penguins's liver, was sublime, the Wincarnis tasted strongly of muscatel grape, and the sweet cocoa was the best drink I have had for nine months.*

The importance of the Midwinter's Day feast in maintaining their mental well-being cannot be overstated. It had sustained them for two weeks beforehand, discussing and anticipating what the menu should contain. And the memory of the treat from the next day forward, when they "once more went back to a subnormal allowance" of food, coupled with the knowledge that "everyday now the sun will come nearer and nearer to us," suggests that life should improve and there is literally going to be sunlight at the end of the tunnel that forms the entrance to their "miserable hole."

Campbell wants to ensure that his men, like successful penguins, make it through their own winter and migration. He had set aside the spare clothes they would need in spring and divvied up the rest between them for living in the snow cave. They have shelter and a modicum of insulation from their clothing. What they most lack, what is most critical, is food.

It is July 10, 1912. They are running low on meat and already down to half rations. The men cook in pairs, with a pair cooking the two meals for a day, then having two days off. Levick cooks with Priestley and it is their day to cook.

Campbell, meanwhile, goes outside at midday for a walk down to the beach where he spots a seal, the first they have seen for three months. Rushing back to the igloo, he grabs Abbott and Browning and they head back down to the ice edge where they find a fat cow and a large bull Weddell seal. Abbott stabs the cow in the heart with his knife, but the bull seal is a harder proposition as it heads for the water. Abbott bashes it with his ice axe to little effect and, as a last resort, jumps upon its back, whacking it on its nose with his ice axe, which stuns it enough to bring it to a stop. He reaches out to Campbell for his knife but he does not notice that Campbell has handed him Browning's knife instead. The handle of Abbott's knife is bound so as to form a stop between the handle and the blade, but Browning's is not. Abbott thrusts the knife with all his might into the thick blubber of the bull seal to penetrate its heart. The handle of Browning's knife is greasy and, without the stop, his hand slides unimpeded down the sharp blade, slicing deeply through the base of the three middle fingers on his right hand.

Campbell sends Abbott back to the igloo while he and the others butcher the two seals. By the time Abbott gets to Levick, "His fur mit (sic) was nearly full of blood which soon froze into a solid block." Levick faces a difficult choice: his hands are "filthy & soaked with blubber from the stove," his fingers are "stiff with cold," and he has only "the guttering light of a blubber lamp held by Priestley." He opts to dress the wounds in bandages right away rather than risk infection by trying to open the wounds up to see if the tendons have been cut, because even if they have been, Levick doubts that he will be able to find the severed ends and repair them in the dim light with his frozen fingers. Yet he frets about his decision:

> *I shall feel rotten about it if his tendons are cut, but think it would have been risking serious suppuration if I had attempted enlarging the wounds and picking up the severed ends, even if I had been able to find them in this light, so great was the filth of my hands & whole surroundings.*

The next evening, Levick washes and dresses Abbott's wounds. He observes, "The tendons of three fingers are cut I am sorry to say."

Two days after finding the two seals, Browning and Dickason find another two and butcher them. Things are looking up on the food front. "We had another double hoosh," Campbell notes, to celebrate their changing fortunes.

Apart from the physical hardships that living in a hole in a snow-bank during an Antarctic winter poses, the psychological hardships are no less of a risk to their chances of surviving the almost impossible. The cold and the perpetual darkness had played havoc with the minds of the crew of the *Belgica* during the first Antarctic winter endured by men—and that was on a dry ship with plenty of food. Inasmuch as Levick is the Northern Party's doctor looking after their physical well-being, he also plays a crucial role in attending to their psychological well-being. He recognizes the need to remain cheerful and positive in outlook. Each night, by the flickering light of a blubber lamp, he reads to the men from one of the few books they have with them. First, Boc-caccio's *Decameron*, followed by the newly published novel *Simon the Jester* by William John Locke, then Charles Dickens's *David Copperfield*, and afterward, Balfour's *Life of Robert Louis Stevenson*.

Decameron is an unusual choice: the focus of the one hundred stories it contains is often on human sexuality and depravities. Levick, ever one to display his Victorian-bred roots, pronounces it "a most boring produc-tion." However, the parallels with his observations of penguins cannot have escaped him: Boccaccio's 14th century Italian tales deal with lust, insatiable sexual arousal, deception, and infidelities—all things he had observed in the penguins. And yet, nowhere in the blubber-stained diaries the men keep in their dark cave, in which they discuss food and life in minute detail, is there a single mention of them discussing the behavior of penguins, and certainly nothing in reference to *Decameron*. Their only references to penguins concern how good they are to eat.

Crucially, Levick becomes Campbell's confidant. Campbell is a leader in that he makes well-considered decisions, but he is not a leader who inspires, not a person given to emotion, and certainly not one to reach out to the men. Even within the confines of the most wretched living conditions imaginable, he still maintains naval disciplines, including booking Browning for misconduct for turning out late from his sleeping bag to cook the meal.

Levick is able to engage Campbell in a way that the others cannot. He and Campbell spend long hours in the darkness discussing in great detail the various motorbike tours they will undertake when getting back to civilization. This inevitably involves "dining sumptuously at the various inns on the way," where they discuss ordering the meals in "the most minute particulars, wine and all."

> *It is uncommonly cheering to think of the stretches of white dusty road at home at the present time, with green trees and flowers, pretty girls in summer dresses, and all the other things there that make life good, including the motor-bike I'm going to buy when I get back, until one feels inclined to smash down the door of this damned dismal little hole and clear out, only there's the beastly thin Plateau nosing round outside. It wont (sic) be like this on the Saskatchewan!*

Indeed, the Saskatchewan. One journey, in particular, exercises Levick's imagination. He decides that he will become a writer, "and one of my hottest ideas at present is a canoe trip from the Rocky Mountains to the Atlantic, right through Canada, down the Saskatchewan, and writing about it, with plenty of good photographs." Campbell, who knows Canada well, helps him by sketching out a map of the journey.

> *Campbell & I spend hours over planning my trip down the Saskatchewan. He knows the country round Winnipeg and Edmonton, and has shown me the route on a rough map.*
>
> *It will make the subject for a good book with fine photographs and should not take more than four months, so that six months half pay would easily see me through it.*
>
> *I write now, to read later, that if I dont (sic) do this trip I ought to be led out and shot.*

Led out and shot? At that moment, there are a heap of other things likely to kill him first, not least being the two hundred miles of crevasses and chasms lying between him and safety.

CHAPTER TWENTY

RETURN JOURNEY

As the men lie in the cave, the thought that most terrifies them about getting out of the snow cave and their march to safety is the Drygalski Ice Tongue, which they know from Mawson's arduous trip to the Magnetic South Pole on Shackleton's Nimrod Expedition is a minefield of crevasses just waiting to suck them down at any time. They do not feel in a condition to cope and it seems like it would be such a cruel shame to survive the winter but not the perilous march to Cape Evans. Originally, they had hoped to be able to sledge to Cape Evans over the sea ice, as they expected the surface of the sea to freeze over during the long cold winter. But their forays out from the cave reveal that the winds and storms have produced mostly open water. Even if some freezes, it will be too unstable and too risky to cross. Their only option is to go across the Drygalski, which Levick confesses makes him apprehensive:

> *Personally I am looking forward to the sledge journey before us with mixed feelings, and I think this applies to all of us: relief at getting away from this dismal squalid life, and a little*

reluctance at the idea of the bad time we are probably in for, crossing the Drygalski Barrier.

The other big decision they face is when to leave. Campbell is desperate to get away as early as possible, but Levick and Priestley much prefer to wait until the temperatures are more conducive to spring sledging.

Campbell says he means to start on or about the 22nd Sept. Priestley and I are both against starting till the end of the first week in October, seeing no reason for starting earlier, and the temperatures will have risen by then.

As it turns out, there is no need for the Antarctic to teach the same lesson to Campbell as it had to Amundsen: in the end, it is severe diarrhea, which affects them all throughout September, that renders them unfit for walking anywhere save for crawling to the roundhouse at frequent and irregular intervals.

The epidemic of diarrhoea continues in spite of precautions. Tonight the hoosh, which had liver in it, was distinctly "gamy" in flavour. All have joined the ranks, Priestley and I holding out longest of all.

Levick starts to keep a scoreboard of poos: "Campbell 2. Levick 2. Priestley 4. Abbott 3. Browning 3. Dickason 2."

On September 5, 1912, Levick needs to rush to the roundhouse, only to have Priestley come shuffling down the shaft "in the last extremity." So they alternate turns at voiding their bowels until Priestley returns to his bag. But Levick is there for forty minutes, according to Campbell, and becomes very cold.

It is September 6, 1912. Levick writes: "Our small stack of literature is disappearing fast." Presumably it is what they use to clean themselves. The image of Levick, hunched over in their tiny toilet area scooped out of the snow, wiping his bum with pages of the *Decameron*, strikes me as a metaphor for the contradictions and tensions apparent in his penguin study: wild sex, perverted even, for which Levick cannot hide his contempt, and disgust.

—∞—

At a certain point in the Antarctic winter, when the nights are still long, even as far north as latitude 67°S northwest of the Balleny Islands where the Adelie penguins have congregated, the penguins receive hormonal signals from their pituitary glands that it is time to head home, time to head back to the breeding grounds. By now they are fat from gorging themselves on krill and fish when they are not resting on pack ice. Even so, the journey they face is a daunting one: for those that breed on Ross Island, it is at least a thousand-mile swim. But such is the biological imperative to breed, such is the power of their hormonal instructions, that they do not stop to think about it once, let alone twice.

—∞—

When the Northern Party emerges from the snow cave on September 30, 1912, having survived the unimaginable, their lives still look remarkably precarious; the odds of their getting back to Cape Evans and safety, extremely low. For seven months they have existed on a diet rationed strictly by Levick. The only time he really relented was for their Midwinter's feast on June 22: "One of the memorable days of our lives."

They have been living in the same clothes for nearly nine months since disembarking from the *Terra Nova* on January 8. Just before leaving the snow cave, they change into what new clothes Campbell stowed away for just this moment, even though it had seemed at times like it would never come. When Priestley takes off his blubber-soaked trousers, they stand up by themselves. The men load their two sledges with seal meat, blubber, and the supplies of sledging biscuits and other items that Levick had demanded that they retain during the long winter months, even when their empty tummies were saying otherwise.

In many ways, Levick regards it as something of a miracle that they have managed to endure the seven months of an Antarctic winter in a snow cave and all are able to stand at the end of it. Yet Dickason, and especially Browning, are too sick from chronic diarrhea to do much more than stand, let alone pull the sledges.

In the end, the Drygalski Ice Tongue proves to be straightforward—at least as straightforward as it can be for six men drained and strained by a winter's meager rations of seal meat and almost a continuous month of diarrhea. They discover a gently sloping snow slope to haul their sledges up, and there are mercifully few crevasses in the area. They get across this obstacle, which had loomed so large in the darkness of the igloo, in just three days. At last, it really starts to seem to them that salvation may be possible.

There are still obstacles to face, of course, such as running out of food. Yet, each time they are in danger of that, they discover a seal they can butcher. Eventually, they arrive at Butter Point and the food depot left by Atkinson, gorging themselves on food such as they had never thought they might see again. After that, on November 3, one of their sledges collapses and three days later, close to Hut Point, their remaining sledge gives out. They are able to "borrow" a sledge from Scott's old hut from the Discovery Expedition, which ever since has functioned more as a sanctuary than a storage facility.

Finally, on November 7, 1912, all six men of the Northern Party arrive at the Cape Evans hut, having rescued themselves. The hut is practically deserted. The cook and the physicist, Frank Debenham, are the only ones there. They are shocked and turn pale. "I believe they thought we were ghosts," wrote Levick afterward. And in a way, they are. They have come back from an almost certain death.

Everyone else is out looking for the luckless Scott, who they know for certain is dead.

It is November 12, 1912. Led by Atkinson, the search party had set out near the end of October with two dog teams and mules, which had been brought down by the *Terra Nova*. Apsley Cherry-Garrard is driving one of the dog teams some eleven miles south of One Ton Depot when he sees Charles Wright, who is out in front and acting as navigator, swerve to the right in a spray of snow. Wright has seen a small object and has gone to investigate. It turns out to be the top few inches of Scott's tent sticking out of a six-foot high drift of snow. Wright motions the rest of the party over and then walks up to Cherry-Garrard, saying simply, "It is the tent."

They find the entrance to the tent but it is too dark to see inside. It is only when they clear away the snow from the walls of the tent that the grim scene it houses becomes apparent. Birdie Bowers is lying with his feet closest to the entrance. In the middle there is Scott lying on his side with his sleeping bag half undone and an arm flung over Bill Wilson. Their skin is a patchwork of glassy yellow and frostbite. As Cherry-Garrard puts it when describing seeing his two dear friends and Scott—three men he potentially could have saved—in this tent, so close to where he had waited for six days in March without going further south, seems to attest:

That scene can never leave my memory.

Atkinson, as the leader, takes it upon himself to remove the three men's diaries and letters. There is a loud crack "like a shot being fired" as he moves Scott's arm to reach his diary. The arm breaks. With Wilson is Cherry-Garrard's beloved green-bound copy of Tennyson's *In Memoriam*, which he had lent to Wilson the previous summer when he had said goodbye to the Polar Party at the top of the Beardmore, as they marched on toward the South Pole.

They collapse the tent on the three bodies and build a twelve-foot-high cairn of snow and ice over it. On the top, in a touch of deep irony, the Norwegian skiing expert Tryggve Gran fashions a cross using his own pair of skis. Inasmuch as they mark the fallen, they also tell much of the reason for the fall.

As the search party heads back to Cape Evans, Atkinson, Cherry-Garrard, and the dog-handler, Demitri Gerov, go ahead. On November 25, 1912, they arrive at the Hut Point hut. Cherry-Garrard finds a letter from Campbell tacked to the door by Levick, who had skied across from Cape Evans with it. The news of the safe arrival of the Northern Party is the one bright spot for Cherry-Garrard in what are pretty bleak times:

> It is the happiest day for nearly a year—almost the only happy one.

The long-awaited reunion of the Northern Party with their fellow members of the Terra Nova Expedition is muted by the details of the demise of the Polar Party. According to Priestley, when the Northern Party arrived at Cape Evans, "We were entirely free from fat, and, indeed, were so lean that our legs and arms were corrugated." During the weeks since their return to the hut, Levick and the others who had eked out the barest of livings in the snow cave, have eaten expansively. Everyone is amazed at what good condition they seem to be in. Levick, especially, has put on so much weight he is likened to be the spitting image of King Henry VIII: it seems that when he went through that gate at St Bartholomew's he hadn't quite left it all behind after all.

—∞—

It is November 12, 1912. Commonwealth Bay. Five days after one heroic journey ends and on the same day there is confirmation of the tragic end to another, a different journey is just beginning. After a breakfast of penguin omelets, Douglas Mawson and his two companions, Xavier Mertz and Belgrave Ninnis, are setting out from their base at Cape Denison to get as far east as they can, to try to explore the land he had originally planned to explore from Cape Adare.

It is a difficult journey that takes them high up toward the plateau, across crevasse-laden glaciers. By December 14, they have managed to cover 315 miles, but some of the dogs are struggling. They have divided the dogs so that the six strongest are pulling Ninnis's sledge, which contains their tent, the dogs' food, and most of their food. The six weakest dogs are with Mawson's sledge, which contains some spare food, equipment, and cooking fuel. Mertz, an expert skier, is out front scouting their route across what appears to be a benign surface compared to the nightmarish conditions that they have struggled through so far. At a certain point, Mertz signals to Mawson, who is coming up behind him. When Mawson reaches the spot, at first he does not see anything untoward, so he proceeds, but spying the faint telltale marks of a crevasse lip filled with snow, he turns to signal Ninnis, who is coming up the rear. Mawson then carries on following in Mertz's path.

As Mawson describes it:

> *There was no sound from behind except a faint, plaintive whine from one of the dogs . . .*

He imagines that Ninnis is whipping one of the dogs, but, when he turns again, there is no sign of Ninnis, the sledge, or the dogs. Mawson and Mertz rush back to discover "a gaping hole in the surface about eleven feet wide." Mawson leans over the edge and shouts into the dark depths below.

> *No sound came back but the moaning of a dog, caught on a shelf just visible one hundred and fifty feet below.*

After some hours "stricken dumb" and calling forlornly down the crevasse, the true awfulness of their own situation hits them like an avalanche. They have only one-and-a-half week's rations for themselves and nothing at all for the dogs. They have no tent. They are over three hundred miles from their base, from which it has taken them just over a month to get here, with full rations and fresh dogs.

With so little food, the prospects that they will die of starvation, if the crevasses do not get them first, seems much more than a probability. It is a dead certainty. Perhaps one of them alone might stand, as Scott would have put it, "a dog's chance," but there is no way there is enough food for both of them to get back to the hut.

Their situation is not unlike that faced almost every season by Adelie penguins in the race to fledge their chicks. Is there enough food to feed two chicks? And, if not, how should it be distributed when there is a brood of two chicks?

It is January 15, 2005. My son Daniel and I, together with an assistant, fly into Cape Bird. The massive bergs B-15A and C-16 are still stuck fast on the northeast side of Ross Island. From the helicopter, B-15A looks like another ice shelf, not just a piece of one: it goes for as far as my eye can see, a white, flat-topped wall sitting on the horizon. There is little open water in front of the colony. The pack ice is jammed up against itself. A pod of killer whales moves from one small open lead of water to the next as it passes by, but most of the penguins, both approaching and leaving the colony, are walking in lines across the white expanse of pack ice.

The colony itself looks like a war zone—a killing field. Everywhere there are the carcasses of chicks. Chicks that have died of starvation this year; chicks that died of starvation the previous year; and chicks from the years before that. Since 2000, when the giant icebergs came to a crunching stop in front of the penguin colonies on Ross Island, blocking their normally easy access to open water and food, breeding success for these colonies has varied from "bugger all" to "bugger this." In the cold, dry Antarctic air, the emaciated bodies of the dead chicks have been freeze-dried, flattening them further, like dry bits of cardboard. They are too many and too unattractive for even the skuas to bother with them.

I notice that there are far fewer chicks than normal in the colony for this time in the season, and way more adult birds. Many parents had deserted their eggs before they could even produce chicks, resulting in many more failed breeders showing up during the Reoccupation Period.

If the parent birds have just a single chick left to look after at this stage when the chicks start crèching, their game plan is pretty simple: bring back as much krill and fish as fast as they can manage for their chick. But parents that have two chicks, in years when food is so difficult to fetch, face a curious choice: should they attempt to feed both equally and almost certainly be unsuccessful, or should they favor one chick at the expense of the other, and have at least "a penguin's chance" of fledging it successfully?

Even more curiously, Adelie penguins have at their disposal a potential mechanism for meting out the dinners differently to their offspring. They are one of only a few species of penguins that engage in what are known as feeding chases. When chicks are crèching, typically a parent with a full belly returns to the subcolony and trumpets loudly. The chicks can recognize their parent's call and so they come running over, sometimes with a gaggle of other, unrelated, chicks in tow. However, instead of feeding its begging chicks as might seem like the first duty of any responsible parent, they turn and scamper away, followed by the chicks. Any unrelated chicks soon lose interest and fall away. That does not stop the parent bird. It will often continue to run, sometimes for several hundred yards and oftentimes outside the subcolony, before stopping to feed one chick and before turning and running off again.

Crèche-age chicks look like koala bears with no ears and beer bellies: they are pear-shaped conglomerations of gray fluff, standing up to a foot tall, with two floppy flippers hanging at their sides. By themselves, they are as ineffectual at protecting themselves against skuas as a puff of wind would be. It is their collective mass in a crèche surrounded by the unemployed adults that protects them. Why, oh, why then, would a parent run out into the no-man's-land between the subcolonies and risk having its chicks pounced upon by opportunistic skuas? Among the first things that Daniel and I observe is that, indeed, quite a few of these chasing chicks get attacked by skuas and then brutally killed. Typically, the skuas work in pairs, taking up to thirty minutes to kill a chick by tugging on either side and literally pulling it apart with beaks adapted for feeding on fish, not miniature koala bears.

This obvious potential cost of feeding chases to the parents' breeding success suggests that there must be some advantage to the parents that comes from engaging in this weird form of food distribution. The most favored explanation proposed that it provided the parents with a mechanism to favor one of their kids if food was tight. As the eggs of Adelie penguins are laid three days apart, even though the first egg is only partially incubated until the second is laid, the chick from the first-laid egg typically hatches a day or more before its younger sibling. This gives it a head start in the race to grow ahead of its little brother or sister. By engaging in a feeding chase, so the argument goes, the bigger one will be able to outrun and outmuscle its smaller sibling, meaning it gets fed first. In times of plenty, there should be enough food left for the little squirt, but when times are hard, the big 'un gets virtually all the spoils. The scientists needed a suitable name for this so, of course, instead of calling it favoritism, they came up with the multisyllabic "facultative brood reduction." It just means that, theoretically, the parents can adjust the size of their brood to the available food supply, condemning one of their chicks to become freeze-dried shadows of their former selves in years where the household income of krill is insufficient to rear both offspring.

It is a nice idea to explain what appears to be a not so nice behavior, except that, like a lot of theories in science, it is probably wrong. I had studied feeding chases during a previous summer at Cape Bird and found the complete opposite to this so-called brood reduction hypothesis. Rather than distributing food unequally to their chicks, I found that feeding chases helped parents distribute the food equally, counteracting the bullying get-out-of-my-way behavior of the biggest chick to its younger sibling. Even though the big first chick (the Mawson, if you like) tended to get the first feed by turning and running, the big chick often got separated from its parent, allowing the little chick (the Mertz, if you will) to get its fill. Let's call it the "brood maximization" hypothesis, for want of a shorter word.

The problem with really deciding between these two competing hypotheses is that they both predict the same things when food is plentiful: that is, both chicks are going to get enough food. It is only when food is in short supply and there is not enough to feed both chicks that the differences can really be determined. In those circumstances, if the brood reduction theory is right, then the oldest, biggest chick should grow at the expense of the second smallest chick, which should likely starve to death. However, if feeding chases function to distribute food equally to both chicks, then when food is scarce, both should get fed, but grow at very much reduced rates.

The trouble with devising an experiment to test this is that no ethics committee in the world, with the possible exception of one with Per Savio on it, is ever going to let you deliberately deprive penguins of food to test this. That is where nature, even at its most terrifying, can be quite wonderful too: B-15A and C-16 provide a natural experiment, requiring parents to travel long distances over the ice and limiting the amount and frequency of dinners they can bring to their chicks.

The first thing we note is that parents with a single chick do not run far or much at all. That is, feeding chases really do seem to be about distributing the food when parents have two chicks: it is then that chases are likely to be prolonged and repeated many times. And yes, even in a season where food is as constrained as this one, we find the feeding chases act as mechanism for the parents to switch feeding between their chicks, and the small Mertz chicks do almost as well as the larger Mawson chicks.

One person who might have a hard time accepting that is Cherry-Garrard. While the men wait at Cape Evans for the *Terra Nova* to return and pick them up, they have an unexpected bonus: a trip to Cape Royds. Cherry-Garrard wants to study the Adelie penguins and collect a series of penguin embryos from the colony there. Oddly, for the world's first penguin biologist, Levick opts to forgo the opportunity to conduct more observations on the penguins and stays, instead, at the Cape Evans hut to work on his photographs and writing.

That leaves Cherry-Garrard free to give his opinion about the function of feeding chases unfettered by the need for any data or facts. In his view, the feeding chases really are to weed out the runts. "The Adélie penguin has a hard life: the Emperor penguin a horrible one. Why not kill off the unfit right away, before they have had time to breed, almost before they have had time to eat?"

Cherry-Garrard is certainly smitten with the Cape Royds penguin colony in a way that James Murray was not.

> *With bright sunlight, a lop on the sea which splashed and gurgled under the ice-foot, the beautiful mountains all round us, and the penguins nesting at our door . . . What then must it have been to the six men who were just returned from the very Gate of Hell?*

Had he been assistant zoologist on the Nimrod Expedition, there seems little doubt that he would have claimed the crown as the world's first penguin biologist. It is probably just as well he did not. His observations of the Adelie penguins at Cape Royds are no more scientifically accurate than his observations of the Emperor penguins at Cape Crozier. They contrast markedly with those of the man who has, indeed, just returned from the Gate of Hell.

For Priestley, another freshly back from Hell, the real prize offered by going back to Cape Royds is the chance to finally climb Mount Erebus, the mountain that had almost killed him when he endured a blizzard for three days in just his sleeping bag during Shackleton's expedition. In addition to Priestley, Dickason and Abbott are part of the party. Ironically, when they are at five thousand feet, they are able to clearly make out Mount Melbourne in the distance, a marker for Evans Coves and the area that had afforded so much misery for them. They never dreamed throughout their long winter that they might be looking at it now from such relative safety and from so far away. Their bodies are still recovering from the intervening miles and months. In fact, Dickason suffers altitude sickness.

It is December 12, 1912. Priestley, Abbott, the Norwegian Gran, and one other finally stand on the top lip of the crater that is the 12,448-foot active volcano, Mount Erebus; the mountain named after Ross's ship.

—

It is December 14, 1912. An even bigger mountain faces Mawson and Mertz. They repack their sledge, discarding everything they can to lighten the load. They make a "thin soup" for themselves by "boiling up all the old food-bags." The dogs are given "some worn-out fur mitts, finnesko and several spare raw hide strips, all of which they devoured." After a brief burial service for Ninnis at the crevasse edge, they turn westward and march toward a death that waits for them as assuredly as their friend's. It is only a matter of time before three will have become none.

It is not long before they kill the first of their dogs, feeding him to the others and saving some of the meat for themselves. The meat from the dogs, which are starving even more so than the men, is "tough, stringy and without a vestige of fat" to the extent that Mawson says, "it could not be properly chewed." Even so, like Levick before him, Mawson must do the math. There is no point diving into their measly food rations now if they cannot last the distance: they must eke them out, rationing them so that as the men propel themselves forward there is at least a prospect that a little may be left for tomorrow's journey.

They are starving and losing condition now almost as fast as the dogs. Using the tent cover that had been on Mawson's sledge, Mertz has been able to fashion a shelter of sorts: at its peak it is only four feet high and there is just enough room for the both of them, but when it is surrounded by blocks of ice it does shelter them from the almost persistent wind.

Mawson has done the math alright, but he knows the drill:

> On sledging journeys it is usual to apportion all food-stuffs in as nearly even halves as possible.

On December 28, 1912, they kill the last of their dogs, boiling the skull and taking it in turns to eat exactly their own share like well-behaved penguin chicks without the need of parental intervention. No feeding chases necessary.

If there had been one, there is little doubt that Mawson would have won. They are both losing condition, both losing their skin, which is peeling off them in large patches. But Mertz is in a particularly bad way, and although the weather is often so bad to keep them bed-bound in their makeshift tent, it is Mertz's inability to get going that is keeping them from progressing.

Like Oates taking away even the dog's chance that Scott, Wilson, and Bowers might have had, Mertz's deteriorating condition is doing the same for Mawson. Yet still he sticks with him. Mawson's diary entry on January 6, 1913, could have been written by Scott about Oates.

> *A long and wearisome night. If only I could get on; but I must stop with Xavier. He does not appear to be improving and both our chances are going now.*

The next day Mertz is no better and unable to move. Mawson is sounding more like Scott than ever, "this is terrible; I don't mind for myself but for others. I pray to God to help us."

In the early hours of January 8, 1913, Mawson reaches out to his companion to find him stiff. But by now a blizzard is blowing, the food is almost all gone, and he is still about one hundred miles from the hut. The math looks obvious to him:

> *There appeared to be little hope of reaching the Hut. It was easy to sleep on in the bag, and the weather was cruel outside.*

It is three days before he can get going again, but his feet hurt so much he decides to take off his boots to examine them. He is shocked by what he finds: "the thickened skin of the soles had separated in each case as a complete layer." Using bandages, he binds the soles of his feet back in place and puts six pairs of woolen socks over them.

It takes him the best part of another month but, miraculously, on February 8, 1913, Mawson stumbles into the base at Cape Denison. The *Aurora* has just left that morning, but five men have voluntarily stayed behind for another winter in the hopes that they might find the missing Mawson, Mertz, and Ninnis. Using the radio towers at the Cape Denison hut, they send a message to the ship, but the weather deteriorates, preventing it from getting back and Captain Davis has no option but to take the *Aurora* to pick up Mawson's Western Party, leaving Mawson to face another winter in Antarctica, but one with plenty of food.

⁂

At Cape Evans, the men are hoping that their ship will come for them. However, no one knows when, or even if, the *Terra Nova* will return. Campbell, as the senior officer, has assumed command of the whole expedition, and in an act of supreme irony given what the Northern Party thought they had left behind them, on January 17, 1913, he orders the killing of seals and penguins should they need to endure a third winter in the Antarctic. The next morning Cherry-Garrard finds two seals and butchers them, but that afternoon, to the relief of everyone, the small ship is sighted steaming up McMurdo Sound. The following day, the remaining mules and three dogs are shot, the hut is shuttered up, and all board the *Terra Nova* to head further south to Hut Point for one last task: the erection of a memorial to the Polar Party.

The carpenter has fashioned a large twelve-foot cross out of Australian jarrah, a hardwood, and in a procession that must have looked for all the world like that accompanying Jesus up Calvary Hill, seven men from the search party, including Cherry-Garrard, carry the cross to the top of the nearby 754-foot-high Observation Hill. In addition to carving into the cross the names of the five men who had succeeded in marching to the South Pole but died trying to get back, there had been debate about including a quote from the Bible. It was Cherry-Garrard who interceded and insisted that it was most appropriate to use, instead, a line from Tennyson.

When erected, the cross stands nine feet out of the ground and overlooks the barrier ice, which was both a route and a resting place for the five men. Cherry-Garrard picks up a piece of the basalt rock at the base of the cross, a memento for Bill Wilson's wife, Oriana, "Facing out over the Barrier, we gave three cheers and one more."

It is October 18, 1977, and my first night in Antarctica. I am twenty-three, more than a full decade younger than Murray Levick when he got to spend his first night there. I am much closer in age to Apsley Cherry-Garrard, who was just twenty-four when doing the same. And, just as Bill Wilson had taken the young and inexperienced Cherry-Garrard under his wing, an older and much more experienced zoologist named Gordon Grigg looks out for me. At midnight we grab a bottle of scotch and climb Observation Hill, which sits between New Zealand's Scott Base and the US Antarctic base of McMurdo. It is early enough in the summer season for the sun to still set, and the clouds surrounding Mount Discovery to the southwest are a brilliant orange. The cross has weathered but the inscription carved into its crossbar is clearly visible:

CAPN R F SCOTT RN

DR E A WILSON CAPN L E G OATES INSDRGS LT H R BOWERS RIM

PETTY OFFICER E EVANS RN

Below the names, on the trunk of the cross, there is Cherry-Garrard's chosen line from Tennyson:

To strive, to seek, to find, and not to yield.

We lean against the base of the cross and take slugs of whiskey straight from the bottle. It is very cold—perhaps -20°F—but we stay there for a long time, warmed by the whiskey and stilled by the harsh beauty of it all. At this stage of my life, I know the story of the five men who did not return from the South Pole, but I know nothing of the six men from the Northern Party who had observed the cross being carried up there.

———

The *Shokalskiy* arrives at McMurdo and moors next to Scott's hut at Hut Point, exactly as Scott had moored the Discovery in 1902. I am on my fifteenth trip to Antarctica. Once again, I climb Observation Hill at midnight and contemplate the cross. By now I know a lot about the story of Levick, Campbell, Priestley, Abbott, Browning, and Dickason. I cannot help reflecting on how death has made a hero of Scott and the Polar Party in a way that the survival of Levick and the rest of the Northern Party has not.

When the *Shokalskiy* gets to Cape Evans, I go to Scott's Hut. It is impossible to not be moved by it. In the dry, frozen Antarctic environment, the hut is so well preserved it is as if Campbell has just shut its door for the last time before boarding the *Terra Nova*. I think back to my first time at the hut and the thing that most moved me then was the skeleton of one of the dogs, still tethered, still lying where it had been shot outside the hut. The skeleton has since been removed. This time, it is the sight of Scott's bed that leaves me heartbroken. Soft light filters through a small window and, in the hut's dark interior, illuminates his caribou sleeping bag, which is pulled back, as if awaiting his return.

In that pathetic scene there is so much sadness, not just for Scott but for my man Levick too. I am struck by the thought that had Levick not returned, he too would have been a hero and his journals, too, would have been made public at that time. As it was, Levick returned to England as one of the blank faces of the expedition, his incredible story of survival lost in the noise surrounding Scott's death; the contents of his zoological notebook left a secret for over a century.

It is January 26, 1913. On its way out of the Ross Sea, the *Terra Nova* stops at Inexpressible Island in Terra Nova Bay to pick up the Northern Party's depot of remaining geological samples and to allow some of the others to see the snow cave where the six men had endured the unthinkable. Campbell leads a party to the snow cave that includes Wilfred Bruce, who is stunned by what he sees:

Everything jet black & horribly greasy & smelling of blubber.

It is not the image of Antarctica, the big white continent, that Murray Levick had in mind when applying to his old boss, Captain Robert Falcon Scott, to join the Terra Nova Expedition. And neither had the true behavior of the penguins proven to be what he had in mind either.

If he is sad at leaving all this behind, he does not say. It is sadness at leaving behind the bodies of Captain Scott and their other four companions that casts a pall over the mood of all those on board the *Terra Nova* as it steams past Cape Adare and away from Antarctica.

PART FIVE

AFTER
ANTARCTICA

Prostitution

In the morally righteous and sexually repressed world of Queen Victoria's Great Britain, into which Murray Levick was born, prostitution was viewed as the Great Social Evil. In London alone, it was estimated that over 8,600 prostitutes were plying their trade. They were almost exclusively poor or working-class women who turned tricks for middle- and upper-class men. For them, it provided a livelihood and a means of survival at a time when jobs were hard to come by; for the men, it was an opportunity to give vent to hormonally driven behaviors at a time when society was doing its best to put the kibosh on sex.

The earliest references to prostitution date back to the 18th century B.C.E. and it, perhaps deservedly, is often referred to as the "oldest profession." There is an assumption—actually, much more than that, an agreement—that prostitution is a peculiarly human phenomenon. After all, it requires there to be some form of currency that can be exchanged for sexual favors. For us, that could be anything from coins to goats; anything of value, in other words.

It is hard to imagine other animals trading sex for inanimate objects or, even, goats. Least of all, penguins. Indeed, the only objects at all that have any perceived value in colonies of Adelie penguins are stones: the penguins use them to line their nests. The prospect that a female penguin might allow a male to hump her for a price measured in stones is something beyond the ken of even Murray Levick, who has witnessed everything from necrophilia to pedophilia, from self-gratification to rape by penguins. No, when it comes to sexual depravities, even the Victorian Levick knew that we humans are in a class of our own.

He had only to look at the way men fought with each other to realize the limitless bounds of the moral void that accompanies us compared to the other members of the animal kingdom.

CHAPTER TWENTY-ONE
THE DEPRAVITIES OF MEN

Somehow, I had hoped that my trip on the *Shokalskiy* would have revealed more of Murray Levick than it did. It had helped me understand why he did what he did in Antarctica; how the circumstances conspired to make him the world's first penguin biologist. It had helped me realize the extent of the horrific obstacles that he had overcome to just survive and return from Antarctica; what a hero he was in his own right. But it did not help me to understand what he did after that; why he buried the most interesting and novel parts of his penguin data, whether by design or at the request of others.

I am standing in front of a locked gate. A sign on the fence indicates that it is under video surveillance. It is the entrance to Roald Amundsen's property, Uranienborg, on the banks of the Bunnefjorden. I have come here to try and understand how the fates of those men whose lives had become interwoven with Levick's had changed once they returned from Antarctica.

Amundsen had won. He had gotten to the South Pole first. But he was painted as the villain in the piece: the coldhearted, calculating, deceitful clinician who was carried to the pole on the backs of his dogs. It was the story read to me when I was seven years old.

Behind me, on a small rise, there is a statue that tells the same story. It is of Amundsen standing, staring into the distance, fixated on the prize to the exclusion of all else. With him are not his men, not a sledge, not skis, not Sigrid. It is one of his dogs.

It probably did not help his reputation that Amundsen had deliberately deceived Nansen and the king. And it was true that he could be calculating, coldhearted, even. On his way back from Antarctica, during a stopover in South America, Amundsen learned that Sigrid was interested in pursuing their relationship. When back in Norway, he had his friend Herman Gade inform her that he did not wish to see her again.

However, while he tried to keep it hidden from public view, as much as Levick did the sexual escapades of the Adelie penguins, inside Amundsen was a romantic side that Sigrid had unleashed. Roald Amundsen had been in London giving talks soon after Christmas following his triumphant return from Antarctica. There, a young female was attracted to his ecstatic display and a mate-switching took place: the baton was passed to a replacement that outwardly resembled Sigrid. She was young, just twenty-six at the time, beautiful, and perhaps most tellingly, married. Kristine Elisabeth Bennett, somewhat appropriately, went by the name Kiss.

Earlier that year, when Amundsen had published his account of his expedition, entitled simply, *The South Pole*, he had chosen a particularly telling quotation from author Rex Beach to begin chapter two:

> *The deity of success is a woman, and she insists on being won, not courted. You've got to seize her and bear her off, instead of standing under her window with a mandolin.*

Amundsen had begun chapter two in his heart in similar fashion. Yet as much as he tried to seize her and bear her off, Kiss resisted. At one point, he gifted to Kiss, in his will, the property somewhere down the locked driveway before me now.

I look about. There is nobody around. I jump the fence, hoping that no alarms will go off, that no one will be looking too closely at the video monitor.

The house is magnificent: a large, two-storied gray-and-white Swiss chalet nestled upon slabs of shield rock, almost on the banks of the fjord, and surrounded by forest. It is as idyllic a property as I have ever seen. On the only flat area between the front steps and the pier, I can see where he must have erected Framheim's hut that he had tested out so thoroughly with Sigrid's help. To suggest that his property should go to Kiss, this supposedly coldhearted man must have really loved her. He was playing a mandolin under her window.

As I leave, I glance once more at the statue. If the real Amundsen can be so different from the public persona, perhaps the same applies to Levick? Maybe he wasn't the stuck-up Victorian prude that I had assumed on first reading? Of course, Amundsen and Levick were not the only ones conflicted after coming back from Antarctica.

It is January 3, 1913. Hjalmar Johansen, the man who should have gotten to the North Pole with Nansen and the South Pole with Amundsen, but in the end did neither, walks along the snow-encrusted footpaths to Sollis Park in the central part of downtown Oslo. Just over four months earlier, he and all the other members of the Fram Expedition had been honored by King Haakon VII with the South Pole Medal, echoing the way Shackleton's men on the Nimrod Expedition had been awarded the Polar Medal from King Edward VII. By now, however, alcohol has got the better of him in a way that the snow and ice of the polar regions never could. He has become a drunkard and a wife beater; a broken man whose own childhood dreams have been taken from him. Like Eivind Astrup and Bertram Armytage before him, he takes out a gun and shoots himself.

279

It is 8:00 P.M. when I get to Sollis Park on the anniversary of Johansen's suicide. The park is dark, deserted, and covered in a six-inch layer of snow. It is small and triangular-shaped. It feels more like wasteland than it does the last vestige of nature in a city of concrete and bricks.

I stand in the middle of the park and look around. I am brought to my knees, despite the cold and dampness of the snow, for exactly the same reason I had been when viewing Charles Bonner's grave in Port Chalmers: this seems such a sad setting, a place as far removed from the grandeur of polar exploration and national heroes as it is possible to be. Though, in this case, there is an exception.

The park is overlooked by the National Library, a beautifully proportioned, salmon-colored brick building. I look up at its windows and wonder about the books it contains. Perhaps Johansen felt that his story belonged there too, along with those of Nansen and Amundsen? And, just maybe, his final deed was his alcohol-inspired way of getting to his personal pole and the recognition he felt he deserved?

Certainly, failure and recognition are not mutually exclusive.

Under the cover of darkness, in the early hours of February 10, 1913, the *Terra Nova* hoves to off the small coastal port of Oamaru, a bit up the coast from Port Chalmers. Refusing to identify itself to the night watchman, a dingy is sent ashore containing Dr. Atkinson and Lieutenant Pennell, captain of the ship. As soon as they are able in the morning, they send a coded telegraph so that family members may be informed that Robert Falcon Scott and his four companions are dead. In accordance with an agreement with *Central News* to be the sole agent for worldwide rights to the expedition's story, they cable an initial report about the deaths, including in it Scott's message to the public; painful words written nearly a year earlier in the tent that even now still contains his body on the Ross Ice Shelf. Then they get the hell out of there, but deliberately take their time, thirty-six hours, to sail the short distance up to Lyttelton.

The news is broken first in London's *Daily Mail*. The world already knew that Scott had lost the race for the South Pole, now it learns that he, Wilson, Bowers, Oates, and Evans have all failed to return. By the time the *Terra Nova* slips through the entrance to Lyttelton Harbor on the morning of February 12th, flags at half mast, it is as if a great tragedy has occurred that has personally affected all who are there to greet the ship. As Cherry-Garrard describes it:

> *We landed to find the Empire—almost the civilized world—in mourning. It was as though they had lost great friends.*

Despite his failures, the recognition and adulation for Scott has begun in a way that it never had for Johansen. The *Daily Mail* manages to praise Scott while, unwittingly for sure, pointing to the reasons for his failure. It adapts a poem by Tennyson. Not the words that adorn the cross atop Observation Hill in Antarctica, but the words the same poet wrote in response to the death of John Franklin:

> *Not here! The white South has thy bones; and thou*
> *Heroic sailor soul*
> *Art passing on thine happier voyages now*
> *Towards no earthly Pole.*

Franklin: the man who failed to pioneer the Northwest Passage; the man whose ships, the *Erebus* and the *Terror*, would be the first in the Ross Sea before ending up on the bottom of the Arctic Sea; the man who was as ill-prepared for the realities and rigors of polar exploration as Scott had been; and yet—this is the deepest irony—the man who would inspire the childhood dreams of Roald Amundsen, leading to him eventually beating Scott to the South Pole.

Even in death, Hjalmar Johansen, life can sometimes seem unfair.

Coincidentally, the partners of two polar explorers are at sea at about the same time.

Paquita Delprat is in the Indian Ocean, sleeping in her bunk aboard the *Roon*, when a cabin boy delivers a message. It says that her fiancé, Douglas Mawson, has narrowly survived a tragedy in the Antarctic that has claimed the lives of his two companions, Xavier Mertz and Belgrave Ninnis. She is upset but thankful for the news that her Douglas has survived, even if it means that he must spend another winter in Antarctica, and they, another winter apart.

Kathleen Scott is in the Pacific Ocean aboard the RMS *Aorangi*, heading to Port Chalmers for a scheduled rendezvous with her husband, Captain Scott. She had left their now three-year-old son, Peter, with her mother. On her way from England, she had taken time out to assuage the adventurous spirit that had never really been quenched in her since marrying Scott. She traveled right across the United States on trains and horseback. By chance, in New York she had met the explorer Robert Peary in an elevator. He happened to be there for a dinner honoring Amundsen, and Kathleen could not help herself: she snuck a look inside the banquet at the man who had beaten her husband. Amundsen, she wrote in her diary, "looked unspeakably bored."

The captain of the *Aorangi* personally fetches Kathleen to his cabin and hands her a cable. The cable says Scott and his men have perished after reaching the pole.

As she tells the story herself, she responds:

> *Oh, well, never mind! I expected that. Thanks very much. I will go and think about it.*

What she does do is go straight to a ninety-minute Spanish lesson and, later, plays five games of deck golf. As she writes in her diary the same day, "Let me maintain a high, adoring exaltation, and not let the contamination of sorrow touch me."

Kathleen's stoic reaction to the news may, on the one hand, seem as unfathomable as sleeping with the mentor to her husband's rival. On the other, she has always been upfront about her desire to have a hero for a son. She had reasoned that having a hero as his father should be a good start. Perhaps, even, a dead hero?

———◦———

It is February 2013. A year after Douglas Russell had discovered the hitherto censored paper by Murray Levick about the sexual depravities of penguins. Now, another unpublished paper has come to light, this one also censored from the public's gaze for about a century. Except that this one really is just a piece of paper, a note, written in pencil.

When Dr. Edward Atkinson crawled inside Scott's tent, eleven miles south of One Ton Depot, to collect the personal belongings from the three frozen corpses it contained, he had found a note written by Kathleen in pencil, which had been kept by Scott in a pocket close to his heart. In it, she refers to Peter by their pet name for him, Doodles, but mostly she refers to Scott himself:

> *Look you—when you are away South I want you to be sure that if there be a risk to take or leave, you will take it, or if there is a danger for you or another man to face, it will be you who face it, just as much as before you met Doodles and me. Because man dear we can do without you please know for sure we can. God knows I love you more than I thought could be possible but I want you to realize that it wouldn't be your physical life that would profit me and Doodles most. If there's anything you think worth doing at the cost of your life—Do it. We shall only be glad. Do you understand me? How awful if you don't.*

How awful, indeed. But there is little doubt that Scott understood her completely.

When Kathleen finally gets to New Zealand, she is given Scott's diaries. With them is a letter, addressed to her by Scott inasmuch as it says, "To My Widow." He begins, "Dearest darling, we are in a very tight corner now I have my doubts of pulling through," and goes on to plead with her to, "Make the boy interested in natural history if you can." Its ending, unlike his own, is rather abrupt:

I think the best chance has gone we have decided not to kill ourselves but to fight it to the last for that depot but in the fighting there is a painless end so don't worry.

And, just like the last time she has seen him—on a boat in Port Chalmers—the last time he addresses her, there are no kisses. He understood alright.

⸙

It is March 31, 1914, and there are kisses aplenty. After getting back from Antarctica, Douglas Mawson is marrying Paquita in Melbourne. The next day they leave for England, arriving in London on May 3, where they are met at Victoria Station by none other than Sir Ernest Shackleton. The expedition still has debts of several thousand pounds and Mawson needs to somehow raise enough money to pay his creditors. Over dinner with Kathleen Scott on May 21, she offers to make an anonymous payment of £1,000 to Mawson from the royalties received from the publication of Scott's journals, the two-volume set, *Scott's Last Expedition*. She reasons that Mawson is owed this as compensation for Scott having usurped his plans to go to Cape Adare and sending Murray Levick and company there instead. Shackleton also steps in to help in his inimical way, which, like his support four years earlier for Mawson's original plans to go to Cape Adare, arguably benefits himself more than it does Mawson: he buys Mawson's ship the *Aurora* for a new venture he is planning at the bargain-basement price of only £3,200.

Just over a month later, June 29, 1914, Mawson is, like Shackleton before him, knighted by the king, King George V. Except that the contrast could not be greater: the Great Caresser bonded with the Great

Crooner; the family man king and the new family man Mawson. At least, that is the way it all seems.

Despite the years of abstinence that Antarctica imposed, it has not affected Mawson's breeding success: no Reoccupation Period, no practice run, proves necessary for him. At the very moment that the king is telling the new Sir Douglas to arise, one of Paquita's eggs, which has been fertilized by one of Mawson's sperm, is burrowing into her uterus.

Other events are also taking place and developing at a much faster pace. Just the day before, Archduke Franz Ferdinand of Austria and his wife had been assassinated in Sarajevo. Europe is careening down a seemingly unstoppable path toward war.

In little more than a month, the first shots are fired in what will become known as the Great War.

When Levick and the others on board the *Terra Nova* had headed back to civilization, they might have supposed that they were due a well-earned rest. But these are tough times and Great Britain is about to be drawn into a war with Germany that will involve much of the rest of the world.

Yet Scott had always pitched the Terra Nova Expedition as nothing if not a scientific expedition, rather than a straight-out race to the pole. It was one reason—arguably not a good enough one—why Scott, Wilson, and Bowers were still pulling those thirty-five pounds of rock samples with them on their sledge when they reached their final camp. It falls to the expedition's survivors, therefore, to make sure that their scientific findings are written up before going off to fight the Huns.

In March 1914, just as Douglas Mawson is marrying Paquita Delprat on the other side of the world, Murray Levick publishes the first ever book on penguins, *Antarctic Penguins: A Study of Their Social Habits*. It is based upon the observations recorded in his blue-bound Zoological Notes. It is the book I shall take with me to Antarctica sixty-three years later. However, it treats the mating behavior of the penguins as if they are married couples, like Mr. and Mrs. Mawson. Nowhere in it does he mention the sexual escapades of the penguins that he has witnessed in Antarctica. While he is doubtlessly prevented from doing so by the keeper of zoology at the British Museum of Natural History—as Douglas Russell would later discover—Levick is clearly sensitive about the salaciousness of some of his findings too.

Throughout the winter in the snow cave, it would seem, from his diaries and those of the other men, that he did not discuss his observations of penguin sexual behavior with them, despite being cooped up with them in a way that one would have thought a little hanky-panky between adult penguins would have been the least of their worries. And of course, he went to the elaborate ruse of pasting over the most extreme sections of his Zoological Notes with the code using Greek letters that he had learned as a schoolboy.

Levick is only thirty-seven when he publishes *Antarctic Penguins* and he has already survived what one might imagine are enough adventures for a lifetime. Yet he must go into action as part of World War I, and no doubt at times, he and the other members of the Northern Party will have cause for wishing they were back in the snow cave on Inexpressible Island.

George Abbott—big, tough, handsome George, who could deal with any hardship it seemed, including slicing through the tendons on three of his fingers with barely a murmur—had suffered a nervous breakdown on the ship while going back to England. Once there, he is hospitalized in Southampton and discharged from the navy. He subsequently joins the Royal Air Force but dies from pneumonia in 1926, after flying without his helmet and googles, aged forty-six.

Frank Browning continues to serve with the navy throughout the war, and eventually retires in 1922. The year before, when he is thirty-nine, he marries Marjorie Bending, a woman sixteen years his junior. The couple have two children but, perhaps as a consequence of complications arising from his chronic illness and diarrhea when in the snow cave, he too dies young: in 1930, when only forty-eight.

Three months after returning from the Antarctic, Harry Dickason marries his cousin, Lillian Lowton. He continues to serve in the navy, and during the war he spends his time on the HMS *Baralong*, which is responsible for sinking German U-boats and famously, "taking no prisoners." Dickason receives a Distinguished Service Medal for his role but the insinuation of apparent war crimes by the crew of the *Baralong* leads, in part, to the drafting of the Geneva Convention. As adept as he is at war with guns, Dickason has apparently learned little from the Adelie penguins when it comes to the Sperm Wars: he and Lillian have five children but Dickason is said to have been at sea when the fifth is conceived. He resigns from the navy in 1924 and lives for another nineteen years, dying at fifty-nine.

The "officers" fare a lot better than the "men" of the Northern Party when it comes to their longevity. Whereas the men live an average of fifty-one years, the officers live for an average of eighty-two years.

The geologist Raymond Priestley is the longest-lived of them all: he does not die until 1972, when aged eighty-seven. Upon his return to the United Kingdom he, like Levick, rushes a book into print. In his case, it is a version of his Antarctic journals, which he publishes in 1914 under the title *Antarctic Adventure: Scott's Northern Party*. While it goes some small way toward publicizing the heroics of the Northern Party at a time when, despite the war, the world is still talking about Scott, most of the copies are destroyed in a fire when a bomb dropped from a German zeppelin hits the warehouse in which they are stored, muting the impact of Priestley's voice. Ironically, during the war, Priestley works in areas associated with communications. Thereafter, he follows an academic career path, eventually becoming vice-chancellor of Melbourne and Birmingham Universities. Crucially, with Frank Debenham, he cofounds the Scott Polar Research Institute in Cambridge, which becomes a repository for the diaries and memorabilia of polar explorers, including most of those from the Northern Party, which, one day, will be read by me. Soon after returning from the Antarctic on the *Terra Nova*, he had married a New Zealander, Phyllis Boyd, and, in a curious twist, two members of the Terra Nova Expedition, Charles Wright and Griffith Taylor, married his two sisters.

Victor Campbell serves in the war at Gallipoli, for which he is awarded a Distinguished Service Order, and then he adds a slew of other distinctions to that for his exemplary service in the Dover Patrol and as part of the Zeebrugge Raid: an attempt to block German U-boats from using the Belgian port. He retires from the navy in 1923 and three years later his wife, Lillian, divorces him. Their relationship had never really recovered after Campbell's time in Antarctica from the difficulties it had been experiencing before he went south. In another curious twist, thereafter he marries Marit Fabritious, a Norwegian who was the maid of honor to Queen Maud, the wife of the Norwegian King Haakon VII for whom Amundsen had claimed the area around the South Pole. They settle in Newfoundland, Canada, where he dies in 1956, aged eighty-one.

Levick, after going into action as part of the Royal Navy's Grand Fleet in the North Sea, also finds himself at Gallipoli during the war.

Gallipoli is especially important for New Zealanders. Each year, we mark the anniversary of the landings by Allied Forces on the beaches of the Gallipoli Peninsula on April 25, 1915. During the eight months of the campaign, more than 130,000 men died, including 2,779 New Zealanders: a heavy toll for a country consisting of only one million people, at the time.

Levick by then has been promoted to surgeon commander. The caring he showed for the well-being of the other five men in the snow cave is evident again in the way that he looks after the New Zealand and Australian men, the "Colonial troops" as he calls them, who have been wounded in battle.

———

It is July 18, 1915. Murray Levick is on the Royal Navy cruiser the HMS *Bacchante* off Anzac Cove. He takes out his pen and writes again in the same neat flowing script he had used when writing out his Zoological

Notes at Cape Adare. This time he is writing a letter. It is addressed to a "Mr. Beeton." Levick describes how he left his ship, absent without leave, and boarded the H.M.T. *Saturnia*, after a Catholic priest told him the transport ship was full of seven hundred men and virtually no doctors. He found "the ship packed above and below by a mass of unfortunate men, the majority severely wounded."

> *Many of the wounds had not been dressed since they left the field and were crawling with maggots, whilst the stink of these rotting wounds (on) the hot decks below was almost unbearable.*

He stayed for four days, he tells Beeton, "refusing to return to my ship when they sent a boat for me."

> *I was hoping that I would be court martialled, so that I could have an opportunity of making some sort of fuss, as this sort of thing had been going on for months, but nothing happened!*

He notes that the wounded have been treated with "most scandalous neglect," adding that, "The feeling among the Colonial troops at this callous neglect is very strong."

Three weeks later and the *Bacchante* is bombarding Turkish troops and artillery installations in what will become known as the Battle of Chunuk Bair, a hill that is occupied by the Turks. An offensive involving largely New Zealand troops is launched on the evening of August 6. Two days of bitter fighting later, after briefly taking the hill only to lose it again, there are over eight hundred dead New Zealanders and nearly twenty-five hundred wounded, needing medical care from the likes of Murray Levick.

As horrible as his experience of living in a snow cave during the Antarctic winter may have been, it must have paled into something close to a fond memory compared to the horrors of the war to which Levick is exposed, literally, firsthand. A crew member on the *Bacchante* recalls helping Levick "by throwing amputated limbs over the side."

Levick is in the last party to leave Gallipoli. For many New Zealanders and Australians (who suffered over 8,700 casualties of their own at Gallipoli), Levick is not a hero for what he did in Antarctica, and certainly not for what he did for the penguins, but for what he did for them at Gallipoli.

⁓

The full name of the recipient of Levick's letter is the superbly monikered Mayson Moss Beeton, though Levick does not dare to call him by either his first or his second name, even though the man is only eleven years older than he. Beeton has the distinction of being the son of Isabella Mayson, who had married his father, publisher Samuel Beeton, and then published a book in 1861 called *Mrs. Beeton's Book of Household Management*. It became a Victorian publishing sensation, selling over sixty thousand copies in its first year with advice for housewives that reflected a strong dose of Victorian values, especially those of hard work, thrift, cleanliness, and devotion to one's husband. However, the day after giving birth to Mayson Moss Beeton on January 29, 1865, she developed a fever and died a week later, aged only twenty-eight. Ironically, her death was caused most likely by an absence of those Victorian values that she espoused. Samuel, it seems, had contracted syphilis from a prostitute, which he had unwittingly then passed on to Isabella.

Within three years of her death, *Mrs. Beeton's Book of Household Management* had sold nearly two million copies but things did not go well for Samuel after Isabella's death. He was forced to sell his publishing business and died when Mayson was just twelve. Mayson had his mother's writing chops and grew up to become a journalist for the *Daily Mail*, the newspaper that would break the story of Scott's death and begin his transformation into a superhero; but his real acumen, unlike his father's, proved to be in running a business. In 1905, he established a timber mill in Newfoundland, becoming president of the Anglo-Newfoundland Development Company.

During World War I, Beeton has been appointed the director of timber supplies for the British War Office. However, that is not the reason why Murray Levick is writing to him from Gallipoli. Beeton also happens to be the father of Edith Audrey Mayson Beeton, born on July 30, 1890, when Levick was already fourteen.

Audrey, who, like Levick, prefers her second name, has joined the Red Cross during the war and become part of a team specializing in massage and electrotherapy, which just happens to be another specialty of the world's first penguin biologist, Murray Levick. In January 1915, she finds herself working at St. George's Hill Military Hospital in Weybridge. Her autograph book from the time is full of signatures from New Zealand soldiers in her care, and it seems quite possible that some of them have been invalided back to this picturesque sweep of English countryside from the blood-soaked sands of Gallipoli after being treated there by Levick himself. That summer she takes a course in massage therapy for disabled soldiers.

She is twenty-four when she meets Levick. Tall and an outstanding athlete, she is already an English international at lacrosse. Levick is thirty-eight, a veteran of some of the most harrowing experiences imaginable and many that are not; a man focused on fitness, health, endurance, and hardship. She is not a classical beauty: she has an oval mouth that always seems like it is struggling to contain all her teeth, a more than adequate nose, high cheekbones, and dark hair that she keeps cropped quite short and often hidden under a turban. She has broad shoulders and she is strong, both physically and mentally. Neither is he classically handsome, but he is similarly strong, both physically and mentally. Like a pair of almost identical Adelie penguins, they are made for each other. While Levick is deployed on the HMS *Bacchante*, they write to each other and to "Dear Mr. Beeton," who is to become Levick's father-in-law and a man he will be destined to call "Sir" rather than "Mr." or "Mayson," if for no other reason than Beeton, like Mawson and Shackleton, will be knighted.

It is July 18, 1915. At exactly the same time Levick is on the *Bacchante* helping save the lives of soldiers at Gallipoli, Shackleton is on a different ship, the *Endurance*, in Antarctica's Weddell Sea. It is part of his two-pronged plan to make the first crossing of Antarctica: from the Weddell Sea to the South Pole and onto the Ross Sea. They had left England on August 15, 1915, just over a fortnight after the declaration of war. At the outbreak of the war, Shackleton had immediately offered himself and his ships to fight the war, but no lesser figure that Winston Churchill, first lord of the Admiralty, had turned him down and asked him to proceed with his audacious plan. The ship Shackleton had bought off Mawson, the *Aurora*, has taken a second party to the Ross Sea, whose job it will be to lay depots in order for Shackleton's Transantarctic Party to continue on to Ross Island once they have reached the South Pole.

The problem for Shackleton is that the *Endurance* is stuck fast in the winter sea ice and going nowhere. In a deep touch of irony, the ship had been built for the Belgian explorer Adrien de Gerlache, the man whose ship the *Belgica* had previously been frozen in the Antarctic, nearly costing Amundsen his life had it not been for Frederick Cook. Shackleton had renamed the ship *Endurance* after his family motto, *Fortitudine Vincimus*: By Endurance We Conquer. However, unlike the *Fram*, which was designed with a rounded hull so that it would rise up when trapped in ice, the *Endurance* has been built for strength, designed to bash through the ice, and does not have a rounded hull.

Strength alone is not enough for the *Endurance*. The Antarctic ice is much stronger, and as the winter turns into spring, the ship can resist no more: the hull of the *Endurance* is crushed.

It is October 27, 1915. Shackleton gives the order to abandon ship. Shackleton has dug them into a hole of his own making. They camp on the ice for over five months, drifting northward with the pack ice until, on April 8, 1916, the ice floe they are on breaks up. Using the three lifeboats they have taken with them from the ship, they manage to get to Elephant Island a week later: a small hunk of land off the end of the Antarctic Peninsula. It is the first time the men have stood on solid ground for nearly five hundred days, but they are as isolated as it is possible to be, well beyond the line where they might have even a glimmer of reason to expect to be rescued.

Nine days later, Shackleton takes five men with him in one of the lifeboats, the *James Caird*, to which they have rigged a sail. They travel through mountainous seas that Shackleton describes as the largest waves he has seen in all his years at sea, until they finally reach King Haakon Bay on South Georgia on May 10, eight hundred miles from Elephant Island. The problem is that they are on the wrong side of the island to the whaling station at Stromness that is their potential salvation. There is a range of rugged, unclimbed, unexplored mountains between them and Stromness. They are exhausted from the boat journey and have little food and little equipment. Still, taking two men with him, Shackleton really proves this time that by endurance he can conquer. They arrive at the whaling station thirty-six hours later, guided by a mixture of incredible endurance, bravery, luck, and, as Shackleton believes, God:

> *I know that during that long and racking march of thirty-six hours over the unnamed mountains and glaciers it seemed to me often that we were four, not three.*

It is May 16, 1916. Douglas Mawson goes to see Kathleen Scott at her home on Buckingham Palace Road in London to discuss a relief mission for Ernest Shackleton. He had returned to England without Paquita or their baby daughter, Patricia, a couple of months earlier to assist with the war effort by working for the Ministry of Munitions based in Liverpool. However, he has also been appointed to the Admiralty Committee for the relief of Shackleton's Imperial Transantarctic Expedition.

Mawson, like Nansen, had already been taken into Kathleen's orbit at the time her husband did the dirty on him and usurped his plans to go to Cape Adare. But if he was like Pluto then, a dwarf planet far away, he becomes like Mercury now, drawn closer and closer in her gravitational field and warmed by her. He starts by dining at her place, then they go out together dining and dancing, and finally, down to a cottage for the weekend on a white sandy beach at Sandwich in Kent, within earshot of the battles being fought on the other side of the English Channel.

Did she just bow deeply to him or did she lie down in his nest? It is clear from their language to each other that there was romance of some sort in the air. Mawson wrote to her after yet another weekend together at the cottage, "will always be able to enjoy it—certainly when you have forgotten all about it, it will be fresh with me . . . the aftermath of contemplation lingers on." It could have been her rhubarb pie he is discussing, but it just doesn't feel like it. Their liaison lasts for two months, three tops, until Paquita and baby come out from Australia.

When it is time for Mawson to return to Australia, Kathleen hosts a party for Mawson but pointedly does not invite Paquita, who perhaps understandably has taken a strong disliking to Kathleen. Invited to the party, however, is one of Kathleen's other planets: the mighty Jupiter himself, Fridtjof Nansen.

The pack ice is again Shackleton's nemesis. Using ships from first the Falklands, then Uruguay, then Chile, Shackleton tries to get to Elephant Island and is prevented from doing so each time by the thick pack ice. Eventually, on the fourth attempt he gets there, and on August 25, 1916, picks up all his remaining men from Elephant Island. Everyone has been rescued, everyone has survived. No need for the *Daily Mail*, Shackleton is now a bona fide superhero.

Shackleton, Amundsen, and Scott, indisputably the three leading figures of Antarctic exploration, had spun a web of circumstances that had ensnared the life of my Murray Levick and turned him, at least for a time, into a penguin biologist. Levick's snow cave buddy, Raymond Priestley, sums up the differences between the three men this way:

For scientific discovery, give me Scott; for speed and efficiency of travel, give me Amundsen; but when you are in a hopeless situation, when you are seeing no way out, get down on your knees and pray for Shackleton.

Yet Murray Levick is an amalgam of all three. He had been the moral, if not the actual leader, of the Northern Party, the person most crucial to their survival. In that, he was like Shackleton. He was meticulous in his organization, the untrained scientist in him dwelling alongside Amundsen. And he had Scott's stamina and perseverance. The rugby man. The physical fitness nut.

As I continue my quest to find the real Murray Levick, I am struck that he, like Hjalmar Johansen, has not had the recognition he deserved. And what of his penguin studies? Did he go back to them after the war?

CHAPTER TWENTY-TWO
AFTER THE WAR

November 11, 1918. The Allies and Germany sign the armistice that brings an end to "the war to end all wars."

Five days later, Murray Levick marries Audrey Beeton in Christ Church on Broadway near Westminster Abbey and the British Houses of Parliament. It is a beautiful church that, appropriately, had started being built in 1841 at exactly the same time that James Clark Ross was edging the *Erebus* and the *Terror* into the Ross Sea for the first time, only four years into the reign of Queen Victoria. In a sense the church is the very embodiment of Victorian values, with its delightfully elegant and tall spire that has been removed and put to the side of the main body of the church, like a "cock" sitting beside a "hen," as Levick might have put it. The church is the perfect setting for the joining together of two people who themselves embody Victorian values.

There is another connection too: it is where Ernest Shackleton and Emily Dorman had faced each other over fourteen years earlier and also vowed to honor and obey each other until death do them part. To mate for life, no less.

The war has taken a toll on Levick personally, as if it had been one of his own limbs that he had tossed over the side of the *Bacchante*. He had retired from the navy the year after Gallipoli, furloughed on the "grounds of unfitness." Thereafter, he worked in the electrical department at Tooting Military Orthopaedic Hospital in London, where he specialized in electrical therapy. In March 1918, he had even resumed writing about science, except that this time it was not about penguins. He published a paper in the *British Medical Journal* about using electrical stimulation to activate the muscles of soldiers with "trench foot," a condition whereby prolonged exposure of their feet to the damp and cold of trenches caused them to become numb and, if left untreated, begin to rot.

However, surely, now the worst is behind him—the snow cave, the war—and with his marriage, a new beginning, a brighter future beckons?

⸺

I am sitting in Christ Church Gardens in the heart of London. The church where Murray Levick and Audrey Beeton were joined together in holy matrimony was destroyed in the Blitz during the Second World War. All that remains of the church are these manicured gardens with the mown lawns and carefully trimmed trees. The place is overshadowed by a large chaotic statue to the British composer Henry Purcell, although I am inclined to think that its wild unconstrained nature probably more accurately reflects the music man than does the one of Amundsen I have seen outside his home at Uranienborg. As I sit there, watching the pigeons searching for bread crumbs at the base of Purcell's likeness, I am beginning to understand how the war could have overshadowed everything for Levick; how it should take precedence over his penguin work. I can also understand how it would alter his perceptions about how he might best contribute to society. Writing about penguin sex must have seemed so frivolous, insulting even, to the memories of those who had lost their lives, their legs, or their loved ones in the Great War. Is that what they had fought for? So that he could enlighten them about necrophilia, pedophilia, and rape in penguins? No, I can see why, at best, medicine must have seemed a more worthy option and why, at its worst, writing about penguins having sex would have seemed as depraved as the acts he described.

I am sure that Sydney Harmer no longer had to tell Levick to shut up. He had come to that conclusion himself. He had already used his code of Greek letters to cover the up the worst of his Pygoscelis pornography. His Zoological Notes had been locked away, put out of sight until an antiquarian book dealer living in an elegant apartment not too far from where I am sitting on the park bench should acquire them nearly ninety years later. His book about the penguins had been published, so he had satisfied that ambition. He had published his official report on penguins for the Terra Nova Expedition—which was even more anodyne and anthropomorphic than was his book. But he has done his duty: to Scott, to the expedition, to the penguins. After Gallipoli, as far as I can tell, he never has anything to do with penguins again.

<hr />

Given Levick's observations at Cape Adare about the wild sex lives of the penguins, it is interesting that his own attitudes to sex seem to have been quite different. There really is something of the prude attached to his writing and one could be forgiven for thinking that, despite his fascination with the carnal exploits of penguins, he was simultaneously repulsed by them and, perhaps even, that he did not care for sex at all.

It is December 8, 1920. Audrey and Murray Levick's son, Rodney Beeton Murray Levick, is born. Murray Levick has the decency to bury his own name as Rodney's third, rather put the father's name first, which had so irked him about his own name. However, irrespective of what he calls Rodney, it is almost impossible to find any mention of Rodney, this product of his own loins, whom Levick hides from view almost as surely as he does his descriptions of the sexual goings-on of the penguins. Possibly, this is because Rodney is, as the British so quaintly put it, "not the full quid."

It is January 5, 1922. South Georgia. Shackleton is back for yet another expedition in the Antarctic, back at the very place where "the hand of Providence" had lead him to safety nearly eight years earlier. This time it takes him on a different journey. In the early hours of the morning, he has a heart attack and dies. He is just forty-seven.

Seventeen years have passed since Shackleton and Emily said their vows in Christ Church. Now, perhaps mercifully from Emily's perspective, death really does do them apart. At her request, Shackleton is buried in South Georgia. On the one hand, it is a gesture that he should remain, like Scott, where he has fallen, bonded forever with the Antarctic realm that has so consumed him and other explorers of the Heroic Age. There is also an inkling that she does not want him back.

Shackleton had continued his philandering ways, like the protégé of the Great Caresser that he had always been, even before meeting Emily. There was not just Hope Paterson, there was a string of affairs, like he was one of Levick's lusty penguins high on testosterone. In an interview given in the United States, Lady Shackleton states that she does not believe in reading children fairy stories where the marriage ends happily ever after as by doing so, she might encourage girls to believe that marriage is their only option. "How wrong that is," she is reported to have said.

It is March 3, 1922. Wrong or not, all the planets have aligned for Kathleen Scott and only one is left visible. She is getting married again, this time to Edward Hilton Young, the First Baron Kennet. They have a son, Wayland Hilton Young, who will become the Second Baron Kennet and, like his father, a politician. His half-brother, Peter

Scott, the son of Robert Falcon Scott and Kathleen, will grow up to found the World Wildlife Fund. Kathleen undeniably fulfils her first husband's wishes by making Doodles "interested in natural history," but in an equally marvelous manifestation of her influence, Wayland, while still a member of Parliament, will write a book called *Eros Denied*, which is nothing less than a manifesto for the sexual revolution. What Levick has tried to suppress in penguins, the second son of Scott's wife makes a virtue of in humans. Indeed, Wayland would agree with Emily Shackleton: "how wrong that is" that marriage and monogamy should be seen as a girl's only options.

Brenda Ueland, a New York–based writer, seems to epitomize that. By her own count, she will have "three husbands and a hundred lovers." It is 1929, when she, then thirty-seven, meets the sixty-seven-year-old Fridtjof Nansen and he becomes another notch in her belt.

Nansen's behavior too might have been lifted straight from the pages of *Eros Denied*. His wife, Eva, had died of pneumonia at the end of 1907. On January 17, 1919, Nansen married the wife of a neighbor, Sigrun Munthe, with whom he had an affair fourteen years earlier while Eva was still alive. The new relationship was not a happy one. Sigrun was despised by Nansen's children, and by the time Kathleen remarried he was said to be unbearably miserable, despite being awarded the Nobel Peace Prize in the same year for his work on behalf of refugees from the war. Which may explain, in part, Nansen's continued dalliance with young ladies.

However, the big, blond Norwegian with the rakish mustache and piercing blue eyes is particularly taken with Brenda Ueland. He sends her naked photos of himself reclining on a chaise lounge and he sends her love letters.

> *Here from my window in my tower, I see the maidenly birches in their bridal veils against the dark pine wood—there is nothing like the birch in the spring. I do not exactly know why, but it is like you, to me you have the same maidenliness—and the sun is laughing, and the fjord out there is glittering, and existence is beauty!*

I am standing in Nansen's tower in his house at Polhøgda, looking out at those birches and the sun glittering on the blue waters of the fjord. Nansen's home is now an institute, but the redbrick tower that stands, somewhat incongruously, on one corner of the house is off limits. A narrow wooden staircase leads to Nansen's office, which has been kept exactly as he left it. When I walked through the door, it had been just like going into Shackleton's hut at Cape Royds.

The room is no less than a shrine to Nansen. It is dominated by a large wooden desk that looks out through arched windows to the birches and the fjord beyond. On the floor is a big polar bear skin and the back wall is lined by shelves of books, mostly about the Arctic. But it is the top of his desk that fascinates me most. There is the ancient typewriter, the ancient microscope, and the ancient inkwells. Its top is covered with ink stains and the knickknacks from a lifetime spent exploring. I sit in his chair and look out at the birches. This is where he had written to Brenda. My eyes are drawn to the wooden basket of papers and letters that sits at the back of his desk. Leaning against it is a small photo stuck to a piece of cardboard. It is a portrait of a beautiful young woman, her thick dark hair parted down the middle, eyes fixed straight at the camera. It is Eva.

It is hard not to dwell on the contradiction of Nansen writing to his lover with his dead first wife, whom he had betrayed so many times, staring at him.

A year after his affair with Brenda, Nansen, like Shackleton, dies of a heart attack. It is May 13, 1930. He is sixty-eight. King Haakon VII himself comes to see Nansen on his sickbed.

And it strikes me as odd that irrespective of what men like Nansen and Shackleton do in their private lives, they are still revered; still held up as heroes, if not gods, by everyone from paupers to kings. Whereas those like Levick, who take a more moral and righteous path, seem to get no credit for that and not nearly enough credit for what they do achieve either. Both Nansen and Shackleton had a great lust for life. Theirs was a romantic view of the world, where Browning or birches assumed as much importance as reaching any goddamned pole.

Is that what my Levick lacked? A lust for life that was worthy of his penguins?

———

Amundsen never really recovers from deceiving his childhood hero, Nansen. As a way to make amends, he becomes obsessed with his goal of reaching the North Pole. The south is, evidently, not enough. At first, he tries to reinvent Nansen's technique of becoming frozen in the ice and drifting across the Arctic. As he cannot take the *Fram* this time, he commissions another ship, the *Maud*. In July 1918, near the end of the Great War, the *Maud* leaves Kristiania and travels up the Siberian coast. However, the ice freezes early and extensively that year, and the ship is too far away to catch the currents that it needs to drift close to the pole. After a frustrating three winters, Amundsen leaves the expedition to others. By now, he is convinced that the best way to get to the North Pole is to fly there.

Trygvve Gran is of a similar mind. Just five days before Britain declared war on Germany, he had become the first person to fly from Great Britain to Norway. During the war, he stuck with the British who had embraced him in Antarctica—despite Norway declaring itself to be neutral during the war—and flew fighter planes for the Royal Flying Corps, earning the military cross. He is now in Spitsbergen, organizing his own attempt to fly to the North Pole.

But Amundsen is determined not to lose this race to Gran's team either.

It is May 21, 1925. Amundsen and five other men take off from Svalbard in Norway's far north to fly to the North Pole. They are flying two large Dornier Wal aircraft that lumber slowly down the ice before, reluctantly it seems, lumbering slowly into the air. They have twin engines fixed fore and aft to the top of the wing, which is itself held on struts above the cockpit and fuselage. The planes are as awkward to fly as they look. They are named N24 and N25.

Early the next morning they need to set down on the pack ice to top up their fuel tanks from the supplies they are transporting with them, which is when everything goes wrong. N24 is incapacitated and there is no way for N25 to take off: the ice they are on is covered in three feet of snow. It is freezing and the men do not have much food. With shades of Frederick Cook ordering the men of the *Belgica* to construct a channel in the ice to free it, Amundsen orders his men to construct a runway. It takes them over three weeks to shift five hundred tons of snow in order to make the 1,500-foot runway. On June 15, 1925, N25 struggles into the air—the great polar explorer having managed to get out of a hole.

He gets out of another hole too. Kiss, after a decade of demurring, has finally said that she will leave her husband and come to Uranien-borg. Just before leaving in N24 and N25, Amundsen writes to Kiss:

> *Let me first fly to the Pole and back, and then we shall see. One thing gives me peace and makes everything else unimportant; You are well! Little Kiss, you know I love you with all my heart and you know I am working for one thing only, to get you.*

Except that he has deceived her as surely as he deceived Nansen: Kiss cannot be said truthfully to occupy all his heart. Amundsen has already encountered his next mate and the switch is soon to take place. Her name is Bess Magids. She is a stunningly beautiful Canadian, five foot four inches tall, with dark hair and dark eyes. She is younger, again, than Kiss, being twenty-six years Amundsen's junior. Significantly, she is also married. Amundsen clears Kiss out of the way as surely as the snow that had blocked the path of N25. His runway is ready for Bess.

It is May 12, 1926. Amundsen and fifteen others fly an airship, the *Norge*, low over the North Pole and drop the Norwegian, Italian, and American flags on the very spot that Frederick Cook and Robert Peary had claimed to reach. And, just as Peary did to Cook, the American aviator Richard Byrd claims to have flown his aircraft over the North Pole three days before Amundsen. However, examination of his records shows that he falsified them and though it will take more than half a century to determine that Peary did likewise, their false claims mean that Roald Amundsen is not only the first man to have reached the South Pole, he is the first to have reached the North Pole too!

Unlike Sigrid and Kiss before her, Bess does not demur. She is ready to dump her husband and live with Amundsen by the water at Uranienborg.

It is June 17, 1928. The Italian pilot Umberto Nobile, who had been commissioned by Amundsen to fly the airship *Norge* to the North Pole, has crashed on the pack ice when flying another airship, the *Italia*, back from the North Pole. Amundsen joins the rescue effort and boards a seaplane in Tromsø to fly to Svalbard. At 6:00 P.M. a radio message is received from the plane, and after that: silence.

Trygvve Gran leads the search for Amundsen, but neither the plane nor Amundsen is ever seen again.

Time has finally run out for Roald Amundsen, who, truth be told, probably tried to win more by "seizing" than he ever did by "playing the mandolin." He is fifty-five years old.

It is April 21, 1934. Amundsen's childhood playmate, Carsten Borchgrevink, dies in Oslo. After returning from Cape Adare at the turn of the 20th century, having led the first party of men to winter over on the Antarctic continent, Borchgrevink had more or less retired from the limelight. He had been pilloried by Sir Clements Markham's men in the Royal Geographical Society, the champions of Scott, for the unscientific nature of much of his reports.

Nevertheless, Amundsen had remained faithful to his old friend, and when he arrived back on the *Fram* from his successful trip to the South Pole, he was quick to pay tribute to Borchgrevink's pioneering work.

We must acknowledge that in ascending the Barrier, Borch-grevink opened the way to the south and threw aside the greatest obstacle to the expeditions that followed.

In 1930, the Royal Geographical Society finally acknowledged Borchgrevink's contribution to polar exploration and awarded him its Patron's Medal.

And there it is, again. Within six years of each other, the three Norwegians, who had played such a big role in Murray Levick ending up at Cape Adare and becoming the world's first penguin biologist, are dead.

—⚬—

After the birth of Rodney, Levick continues his work on electrical therapy at St Thomas's Hospital in London, where he is medical officer in charge of their electrotherapeutic department, and at the Shepherd's Bush Orthopedic Hospital, where he holds a similar position. In 1923, he is appointed medical director of the Heritage Craft School for Crippled Children in North Chailey, Surrey, about forty miles due south of London.

It is an area of uninspiring redbrick houses and buildings, but Levick finds his work at the school is very inspiring, where he deals with pupils with a complex combination of physical and cognitive issues. He continues his focus on rehabilitation, trying out different therapies to help children afflicted by the likes of tuberculosis. He will remain its director until 1950.

In the meantime, he continues to work elsewhere too, including the Victoria Hospital for Children in Chelsea, where he helps pioneer the use of light therapy. He is as committed now to the value of exercise as he ever was before he went to the Antarctic, and he is instrumental in promoting the use of exercise for rehabilitation. Something that endears him to more people, perhaps, than anything else is that he champions and trains blind people to be used for massage and physiotherapy—which have become important means of treating the traumas caused by war—despite the prejudices against doing so at the time.

~

Unlike Levick's life before going to Antarctica, I discover a number of archives full of material from his life afterward. Much of it stems from his medical work, especially at Chailey, and Audrey's work for the New Zealand Hospital in Weybridge.

Having survived the unthinkable during the winter in the snow cave on Inexpressible Island, I had supposed that there would be a bond between the six of them—Abbott, Browning, Campbell, Dickason, Levick, and Priestley—that forever more would make them, if not as thick as thieves, then inclined for the odd reunion. But that does not seem to be the case. Sure, Levick's address book reveals that he has the contacts of some of those from the *Terra Nova*, but there is little hint that he actively sought them out—or to be honest, they him.

Campbell does keep up a correspondence with Kathleen Scott. Yet of all the *Terra Nova* souls, it seems like Apsley Cherry-Garrard is the one who makes the most effort. Perhaps he clings to the rest of them out of guilt? He continues to be tortured by depression and psychosomatic illnesses that will plague him until his death in 1959; by the thought that he had not done enough for Scott and not nearly enough for Bill and Birdie. To honor Bill, he writes his scientific record of the expedition, much like Levick had, but along the way, it morphs into *The Worst Journey in the World*, the wonderfully moving account of the Terra Nova Expedition that will inspire my childhood dreams and probably those of thousands of others.

He is instrumental in supporting the efforts of Douglas Mawson after the war to have Macquarie Island declared a reserve, and the barbaric business of boiling penguins for their oil, which is still being operated by the New Zealander Joseph Hatch, brought to a stop. Even Baron Walter Rothschild steps into the fray, backing Mawson and Cherry-Garrard. By 1920, Hatch is out of business and the penguins are able to resume their own business of breeding without fear. Mawson thereafter, concentrates on his own business as a geologist—although he does make one further trip to Antarctica—living until October 14, 1958, six months shy of Cherry-Garrard: the survivors of unquestionably two of the worst journeys in the world, cheating death with equal alacrity.

When he isn't being depressed, Cherry, as his friends call him, can be wonderfully entertaining. He is independently wealthy and forever holding parties at his house. Kathleen Scott and her son Peter often stay with him to attend. And there are a succession of beautiful girls that accompany him, though Cherry-Garrard always gives the impression of being a male Adelie penguin that is not quite sure if he wants to step out of the nest and let the female lie down in it.

I am especially struck by one photo I come across. It is 1926. Cherry-Garrard and his then lady friend are visiting the Levicks. The photo is of the four of them on a garden bench. It is also one of the few photos I find that includes the enigma that is Rodney. Levick and Cherry-Garrard are occupying the bench with little Rodney, wearing a singlet, sitting between them. The two ladies are in summer dresses and sitting on the arms of the garden seat. Audrey has opened her mouth and let her teeth escape in a big toothy smile. She looks genuinely happy. Levick has his arms around Rodney in what is the only display of intimacy I have ever seen from him in any photos, either with Rodney or anyone else. He is smiling too. My initial impression is of a wonderfully happy family. But when I look closer I realize that, like many six-year-olds, Rodney is not sitting still: his feet are blurred from swinging his legs and, I see then, from the way Levick's hands are clamped onto Rodney's shoulders, that what he is really doing is trying to restrain Rodney, more muzzle than manifestation of affection.

Cherry-Garrard is not smiling at all. He is snappily dressed and is supremely handsome, but he is looking quizzically to the side, barely amused, while his partner is looking at Rodney, as if she sees what is going on.

I see it too. There is something in the photo, in the empathy it evokes in me for Rodney, that makes me realize that I am getting close to my goal, close to the real Murray Levick.

CHAPTER TWENTY-THREE

THE POLE AT LAST

I t is 1932. Levick establishes the Public Schools Exploring Society, whose mission it is to take boys from public schools—that is to say, in the weird way that the British name things, private schools—to remote wilderness areas to teach them survival skills and physical fitness. Many of the initial trips are led by Levick and are to the wild parts of Newfoundland.

The society still exists today, with a slightly altered and more inclusive name, as the British Exploring Society. They no longer need participants to be boys, no longer need them to be from public schools, and, indeed, no longer need them to be going to school at all. But the mission remains the same: take young people to places where they endure hardship and from this "snow cave" experience of their own, they will become better people for it.

The Society's offices are based with those of the Royal Geographical Society, on the corner of Kensington Gore and Exhibition Road, overlooking the beautiful, green Hyde Park in London. But as I approach, I am most taken with what is overlooking me. Set into a white alcove in the side of the redbrick building is a life size statue of Shackleton all decked out in his Antarctic gear, complete with balaclava and big fur over-mittens hanging by straps from over his shoulders. He hadn't gotten to the South Pole, but he had gotten most of the glory it seemed. Perhaps Johansen had cause for feeling hard done by; and maybe Scott too?

The insides of the building, at least the parts with the British Exploring Society, are less impressive. There are several people jammed at desks in a narrow room that is cluttered with cardboard boxes. The walls are lined with lots of blue notice boards to which are pinned lists, maps, schedules, and whatever else is needed to coordinate expeditions to some of the remotest parts of the planet. The archivist is a lovely man with thinning gray hair and a body shape that suggests to me that it has been a while since he took part in any of the expeditions. Indeed, it has: he was one of Levick's schoolboys and he is the first person I have met who has actually met the real, living, breathing Murray Levick.

Though not as agile as he used to be, metaphorically at least, he bends over backward to provide me with access to their archives. I am particularly interested in the expeditions that the society ran to Newfoundland. I want to know if Levick kept in contact with Campbell and most particularly, if he and Campbell ever made that trip through Canada. And something else. Call it a detective's hunch, a gut feeling, but somehow, I suspect this could be connected with Rodney.

The archivist goes and fetches me Levick's notebooks, the ones he kept on all his expeditions. As I open the first, from the 1934 expedition to Newfoundland, I sit upright with a jolt: it is as if Levick's ghost has just walked into the room. There before me is the same familiar writing, with its tight small letters joined by flowing strokes, along with Levick's almost complete refusal to use apostrophes. Written in pencil, it reminds me very much of the sledging diaries he wrote when in the snow cave. But there is nary a mention of Rodney in the notebook even though I know from the publicly available British passenger lists—and the archivist confirms this—that Rodney and Audrey had accompanied Levick on the ship to and from Newfoundland.

The notebook from 1937 is different. Somewhat. As with all the notebooks, he lists all the boys alphabetically; seventy-seven in this case.

This time, Rodney is listed. He is nearly seventeen, attending Rugby, a public school, and is old enough to be on the expedition as one of the boys in his own right. He is not the youngest in the group. Next to the names of the boys, Levick writes in-depth summaries about the prowess of each one. Every boy, that is, except for two. Next to Rodney's name, there is nothing, as if he is not good enough to warrant a description from his father, with or without Greek letters.

The other boy without a summary also catches my eye. He is listed as Gurney, E. R. (Harrow) 18.2 and next to his name Levick has written only one word: Killed.

This expedition had gone to the area around Trout River, which is part of Gros Morne National Park on the western side of Newfoundland facing the Gulf of Saint Lawrence and the coastlines that Captain Cook had mapped with such adroitness two centuries earlier. It is also within a felled tree or two of where Sir Mayson Moss Beeton had his timber business.

Edward Ralph Gurney, I learn by reading on, fell down a cliff on August 11, 1937, when he and another boy left the camp without permission and tried to climb down the rock face beside a large waterfall. His companion managed to get back to the camp to raise the alarm, but Levick and the others in the search party were unable to find Edward before nightfall. When they found him the next morning, he was dead.

Suddenly, it is as if this ghost that is Levick turns to flesh and blood and is sitting there with me, because I start to see aspects of him more clearly than I ever have before. What hits me hard, almost as hard as the blow to poor Edward's head that ended his life, is the way Levick reacts to all this. The expedition proceeds as if nothing has happened. They go camping and hiking in the forests around Trout River for the next three weeks with Levick writing in his diary and making notes about the boys, save for Rodney and, of course, no mention of Edward. He does, however, write out in his notebook the cable he sends to Edward's parents. It chills me to read it; God knows how they must have felt upon receiving it:

With very great regret & sympathy I have to inform you that your son was accidentally killed yesterday by falling from a

*cliff. He was killed instantly. Am arranging for his funeral at
Corner Brook the nearest Church Town.*
Commander Levick, Trout Lake, Bonne Bay, Newfoundland

Such detachment. So brutally stark.

Edward's body is buried in the small graveyard at Curling, a resi-
dential neighborhood in Corner Brook overlooking the sea: the same
reward for dying young in a fall that had been accorded to Charles
Bonner. But that is where the similarities end. Levick gets a wreath of
spruce made for the grave with the inscription:

TO YOUR MEMORY

FROM YOUR COMPANIONS

OF THE

BSES. 1937.

So stiff and formal. No Tennyson or Browning. More Scott than
Shackleton. That Victorian reserve. Murray Levick is a man who has
looked death in the eye, his own and those of other men, more than
should be bearable in one lifetime. Yet this chills me too. His complete
lack of passion; his complete lack of empathy. To Your Memory From
Your Companions: for fuck's sake Levick, is that the best you can do?
And you call yourself a writer?

———

I am in St John's, Newfoundland. It is not that I am angry at Levick,
more that I am disappointed in him. He has been my unseen guide,
my Sherpa, and yet after the Antarctic and the travails of war, he no
longer seems to be beating a path I would wish to follow. In my heart,
I hope desperately that I will find evidence that he took that canoe trip
down the Saskatchewan River, which he and Campbell had discussed
at such length in the snow cave. That he tried to write the book about
the journey, even if it, like his paper on the sexual behavior of the

penguins, remained unpublished. Because, in a way that I have not appreciated until now, Levick is becoming my hero as much as my mentor, supplanting the handsome but troubled Cherry-Garrard, who knew a good word when he saw one, for sure, but lacked a scientist's sensibilities. It is all about the journey with Cherry-Garrard, not the end result. I want my Levick to be both. To be the wild adventurer and scientist, but to come wrapped in the romance he discovered in the snow cave: his life in two acts. His second act has begun, but it does not seem to be benefiting from his reflection during the interval, during that winter he spent lying in the dark on the floor of the igloo with Campbell by his side.

St John's is a city that seems at risk of tumbling into its own harbor, a line of ships along its edge, the only thing separating its buildings and people from the sea. From atop a headland, I realize this is because the wharves don't stick out perpendicular from the shore; the ships are all tied up in one long line along the shore, bow to stern, like a conga-line of dancing boats.

I head inland to the flat-roofed brown building on the campus of Memorial University of Newfoundland that is its library. It is here that Victor Campbell chose to have his papers and records deposited rather than in Raymond Priestley's Scott Polar Research Institute, which in itself might indicate some antagonism between men who had managed to coexist for so long on so little with barely a bad word between them. Campbell, it seems, did a lot of fishing and hunting once he settled in Newfoundland with Marit, but it is his diaries and notebooks from the Terra Nova Expedition, and especially those written in the snow cave, that once more bring everything powerfully back to me. The paper is stained with black fingerprints and smudges from the seal blubber. The writing is in pencil, the letters neat and at once larger, more rounded, and more spaced than Levick's hand.

Campbell is, surprisingly, a man of two halves, as we like to say in New Zealand when referring to a rugby game. His writing is to the point, concise, not a wasted word on emotion, not an adjective to describe something as beautiful or horrific, whichever it might be, and surely he experienced both. But his illustrations, sketches in pencil, are wonderfully artful. What the man could not say with words, he said with lines and shading. One, in particular, is especially arresting. It is of the *Terra Nova* meeting the *Fram* at the Bay of Whales, that fateful moment when the lives of the Eastern Party were changed for good and Murray Levick was sent down an unexpected path to becoming the world's first penguin biologist.

Yet, there is nothing in all the papers and notebooks I peruse to suggest that Campbell ever saw Levick again, let alone floated down some river with him. I find that strange, given what they have been through, given their plans, given their shared experiences of the awfulness of an Antarctic winter with very little and the tragedy of a Gallipoli campaign that achieved very little. Especially since Levick's father-in-law lived in Newfoundland and the Levicks came, on numerous occasions, to Newfoundland with the exploring society. How hard would it have been to stop in and see an old chum, Levick, one with whom you had been to Hell and back?

I leave St John's and head into the interior, trying to at least come to terms with what Newfoundland might have offered Levick that it could pull so hard on him compared to Campbell.

Immediately, I identify with the sense of wilderness that would have made Newfoundland a sort of green version of Antarctica for him: the flat barrier outcrops and the trees that go on for miles. I pass through Terra Nova National Park and, illogically perhaps, I assume that it must have been named after Scott's ship. In fact, it is the old name for Newfoundland: Terra Nova, New Land.

Heading further northwest, toward Gros Morne National Park, I really see why this wild country, with its rugged forested tablelands, should be where Levick came. In his worldview, its extremes can push the body and turn boys to men. Except the luckless Edward, who it pushes too far, and the boy who doesn't merit a mention, Rodney, who maybe it pushes too hard?

I find one more photo of Levick with Rodney. This one is of the two of them sitting by the little garden pond at their home, the White Barn, in Old Oxted in Surrey. Rodney has his mother's mouth, oval and bursting with teeth, his dad's ears, and his mum's dark wavy hair. At eighteen, he still looks slighter and smaller than Levick, who is then sixty-two.

In some ways it must have been tough to have Levick for a dad; this man who only held you to restrain you; this man for whom discipline was such an asset as a scientist but such an impediment as a parent.

I cannot help having a soft spot for Rodney. He may have had some issues, but he was certainly not the fool that others made out. No doubt his father pushes him into Officer Cadet Training, a sort of version of the *Ganges*. The Second World War is just beginning—so much for the first ending all wars—and Rodney graduates on March 9, 1940, to become second lieutenant in the Corps of Royal Engineers. Eighteen months later, he is promoted to lieutenant, not long after London has endured a terrible bombing campaign and Christ Church is no more.

During the war, Levick is recalled to the Royal Navy to train a group of commandos to undertake a daring mission called Operation Tracer, which would have seen them sealed in caves beneath the Rock of Gibraltar for months, in conditions reminiscent of his time in the snow cave, to spy on German shipping movements. The mission is never deployed but Levick is: in 1943, he is used to train commandos and special operations forces at Lochailort Castle. There he teaches survival techniques and advises the men on how they can become tougher, both physically and psychologically. He even publishes his lectures as a pamphlet entitled *Hardening of Commando Troops for Warfare*.

Yes, I can imagine that he would not have been the easiest of dads to have.

———

I go to Budleigh Salterton on the coast of Devon where Levick spent the last years of his life. The countryside on the outskirts of the town is as bucolic as can be, with mown fields bordered by oaks and tall poplar trees. A soft light illuminates a low mist that is hugging the ground—ethereal—the remnants of the morning's fog. It seems as if the earth itself is breathing.

At the seaside, the beach is a long gentle curve; a swathe of pale stones upon which a collection of colorful dinghies and small yachts have been left beached like wooden versions of Weddell seals. The sea is flat and calm, barely moving. In patches it is turquoise, in others purple, reflecting puffy bruised clouds that appear more moody than threatening. It reminds me of my last visit to the Cape Royds hut. The sky could have been the very same one.

Budleigh Salterton is a quiet town of restrained beauty, known locally as "God's Waiting Room" because the average age of the residents is supposedly over seventy. Murray Levick had his appointment with God fourteen miles away, at the Poltimore House Nursing Home, where he died on May 30, 1956, from prostate cancer, just weeks before his eightieth birthday. Audrey, fourteen years his junior, lived for another twenty-four years in Budleigh Salterton. There are people in Budleigh Salterton who still remember the Levicks. One doctor says that Audrey was a formidable woman, who demanded house calls rather than coming down to the doctor's surgery, as was the requirement for everyone else: he recalls that he felt like "running a mile" whenever she called.

In a sense, that would not be at all inappropriate: Murray Levick, as I have repeatedly discovered, was a stern disciplinarian who put great stock on physical fitness.

I make my way to the country lane where the Levicks lived. Their house is grander than most, just shy of a mansion. It is a large, two-storied, white roughcast house with a gray slate-tiled roof. Almost all the windows are divided into eight small panes making it look classic, old, and English, all at once. There are plantings all around the base of the house, including a large hydrangea bush by the entrance, which, jarringly, has flowers of different colors: some blue, some purple, some pink.

Rodney had continued to live here even after his mother died. Eventually, he died on March 28, 1999. According to the neighbors, he was "as mad as a meat axe." He lived alone in a single room above the garage while the rest of the house fell into wrack and ruin, the pipes burst, and much of its contents, including his father's mementoes from Antarctica, were damaged. He would often run around the property naked and could be seen timing himself as he biked, repeatedly, the several miles from the house to the beach, always intent on bettering the time he had taken on the previous trip.

From the house, it is possible to look across the pale fields, which are bordered by dark green hedgerows and trees, and make out the blue strip of water on the horizon that is Lyme Bay. It seems much too far to be enjoyable to bike it time and again. I am struck by a vision, probably an apocryphal one, but it will not go away: a naked Rodney hunched over his bicycle as he cycles from house to beach and back again, all the while looking at his wristwatch. I see a son whose sexuality might be regarded as seemingly as aberrant as any of the penguins Levick studied and one might suppose, a son who has never quite managed to live in the shadow of his father's achievements at endurance.

I get access to Levick's house. The people who purchased it off Rodney's estate are still there. While much of the estate was sold at auction, including Levick's Zoological Notes to an antiquarian book collector in London, there is still much that has been left behind in the house. The owners bring out Levick's skis, his initials carved into the wood: the very skis, I believe, that he used to ski across to Hut Point with Campbell's letter for Atkinson declaring that the entire Northern Party had survived the winter in a snow cave and their journey to Cape Evans. And there are many of Levick's framed photos of the Adelie penguins at Cape Adare: his private collection. Had they been hung with pride on the living room walls, I cannot help wondering, or sequestered away like a stack of pornography?

Most intriguing of all is Levick's actual chair, low-backed and covered in cream upholstery. The owners still have it in a corner of the living room. I take myself over to it immediately and sit in it. I feel connected to Levick in a palpably physical way. The chair is extraordinarily uncomfortable, which I suppose is fitting for a man who prided himself on enduring hardship. I realize, in that instant, that in some ways Rodney and I had both been living in Levick's shadow, desperately wanting to move beyond it.

In one way, I suppose, I have. I have discovered an aspect of the penguins' sexuality that Murray Levick would need a whole new Greek alphabet to record, if indeed he could even bring himself to do that. I have discovered not just penguin perverts, but penguin prostitutes.

It is the mid 1990s. Over four seasons at Cape Bird, Fiona Hunter and I study the copulations of Adelie penguins. We are there like foreign war correspondents, covering the Sperm Wars. We are primarily motivated by my earlier work, which has shown that these penguins often switch mates during the courtship period, opening up the possibility that some males risk rearing chicks that are not their own, given that we know sperm can survive for several days in the reproductive tracts of females. Theoretically, at least, when mate-switches occur, the eggs of the female may get fertilized by the male she has been with, rather than her new partner, even though he is the one that will provide the parental care. Our research has shown that these new males have a counterstrategy that involves bombarding their females relentlessly with sperm like the HMS *Bacchante* firing at Turkish positions on Gallipoli.

Like any good war correspondents then, we are concentrating on the conflict, on the guns going off. We are recording every copulation. When did it occur? Who was it between? Did he hit the target? We swab the females to measure the scale of the hits and we induce the males to engage in target practice by putting out decoys: corpses and fluffy toys.

No one—not Levick, not anyone—has ever monitored copulations in penguins as intensely as we do over those four summers. And, like war crimes, there are things you see that you are not looking for, that you have never supposed in your wildest dreams could happen, but the evidence keeps adding up and telling you otherwise.

Because we are trying to monitor so many birds all at once, we really focus on only one thing: the transfer of sperm from the male to the female, the thing that in the pornography industry is called the "cum shot." We are using a recognized method for sampling behavior known as All Occurrence Sampling, whereby we record every instance of the one thing in all individuals to the exclusion of all else. We are concentrating on the cum shot, not the courtship behavior leading up to it.

However, after the first egg is laid, it takes about three days before the female can lay the second egg. During this time, it is mainly the female's male partner that sits on the egg (even though Levick thought otherwise). While he gets incubation underway, the female will often go in search of stones to shore up their nest. The problem she faces is that at this stage in the season, all the easy pickings with regard to nearby stones have long been scooped up by other birds.

We notice that the females adopt different strategies for getting stones. The most industrious and virtuous, those that adhere to the penguin equivalent of *Mrs. Beeton's Book of Household Management*, go searching for stones outside their subcolony, even if it means walking quite some distance. Others adopt a sneaky and arguably sinful approach: they steal stones from their neighbors' nests. All this we know. We have seen it before, even when not covering the Sperm Wars. Levick had seen it too. We do not need to document it again. It is outside of our correspondents' mandate.

However, we start noticing, without looking for it, that some females adopt a different ploy whereby they dupe males out of their stones by pretending they might have sex with them. What these females do is approach a single unattached male, who, with nothing to do but collect stones, has usually amassed an impressive collection of stones that form his empty nest. The female approaches such a male by adopting the preliminary pose for courtship. She bows deeply, bill nearly touching the ground. The male, will typically respond quickly—an anthropomorphic Levick might even say, enthusiastically—by bowing equally low and moving out of his nest. This is the final prelude to copulation, as it allows the female to lie down in his nest before he mounts her.

Except that, what these duplicitous females do is that they don't assume the position, they use the opportunity when presented with the male's vacant nest and his wealth of stones, to reach across, remove one of his stones, and then carry it back to her own nest. The duped male is left horny and down a stone for his troubles. So pathetically desperate are these unpaired males to mate—and well they might be, as the breeding season is disappearing apace and with it any chances for their own breeding success—the female has only to repeat the charade and the male will stand aside again. In one case, we see a female take sixty-two stones from a male using this cloaca-teasing behavior.

We have witnessed an unexpected war crime of sorts. It looks like theft under the guise of sex. Hence, we start to concentrate our attention on what is happening with these sexy crimes, expecting to document just how dumb the males can be. Being duped sixty-two times seems like a lot, but maybe it can get worse than that: the *Baralong* of penguin wars?

What we observe leaves us gobsmacked. On ten occasions, the female, when approaching the male like this, does get in his nest, does let him fuck her, and then she picks up one of his stones and takes it back to her nest, where her actual partner—the "cock" that is going to be looking after the eggs she produces—is sitting on her first-laid egg none the wiser.

Stones really are the currency of the colony, the things that Levick has shown convincingly keep the eggs free of meltwater. In the penguins' world, they are worth something. To them, they are really the equivalent of dollars or goats. The exchange of sex for a stone is, surely then, prostitution by another name.

Perhaps the males that are being duped are not so stupid after all? There is the possibility, however faint, of feathered fornication. And, all for a price measured in stones.

Prostitution is often said to be the "oldest profession." We humans and our near ancestors have been around for about six million years at most, but, given that penguins have been around for sixty million years, it seems there really is some truth to that.

While our observations may not have shocked Samuel Beeton, they definitely would have poleaxed Murray Levick. Even if he did not consider these war crimes, I know him well enough by now to be certain that he would have regarded the behavior of the penguins as criminal.

—◦◦◦—

I continue to sit in Levick's chair, taking in everything I have learned about Murray Levick. He is a far more complex individual than the cardboard cutout censored scientist I had started out investigating. There is so much to admire about him, but I realize I have become invested in wanting him to be even better than that, to be my inspiration, my new Cherry-Garrard.

He does not need to be Cherry exactly. I want him to have, instead, written *The Best Journey in the World*: that canoe trip down the Saskatchewan, from the Rocky Mountains, through Edmonton, past the window of my office in the Biological Sciences Building where I had analyzed my penguin data from that first summer I had spent studying the penguins at Cape Bird, past Owen Beattie's office where he analyzed his data from autopsying Franklin's men, right down to Lake Winnipeg.

I write now, to read later, that if I dont do this trip I ought to be led out and shot.

Well, Murray, did you? I have been through Campbell's diaries and papers deposited in the Memorial University of Newfoundland. I have re-read through Levick's prodigious writings found in Cambridge at the Scott Polar Research Institute, and in London at the British Exploring Society, and in a book of Zoological Notes in a drawing room in Kensington. I have been through every other archive I can find that contains even a morsel of Levick's life. And nowhere is there even a hint that Levick or Campbell ever did make that trip across Canada that so occupied their minds in the snow cave, either down the Saskatchewan River or on motor bikes.

Then, quite by chance, the archivist at the British Exploring Society had informed me about a letter written by a retired naval man in Vancouver. It says that Murray Levick arrived on his doorstep having ridden a bike right across Canada. Except that, it cannot have been Murray Levick: this was the 1950s, Levick was near eighty, near death. It could only have been his bike-riding son, Rodney Beeton Murray Levick, fulfilling his father's wish, moving beyond his father's shadow in the only way that he could.

Sitting in Levick's chair, looking around his room, I feel a sense of freedom. He is no longer my Amundsen, he is simply my "father" as a penguin biologist, the person who, wittingly or not, has nurtured my interest in penguins and their sexual behavior. Were he here, sitting in the room with me, puffing slowly on his pipe, I hope he would have taken a quiet sense of satisfaction in my own discoveries about the remarkable sexual activities of penguins, in much the same way I hope he would have appreciated Rodney's achievement.

In the end, Murray Levick has shown himself to me in ways that I could not have imagined at the outset of this journey. His meticulous discipline, a product of his Victorian roots stretching as far back as his initial upbringing in Newcastle upon Tyne, predisposed him to be a great observational scientist even without any zoological training. It predisposed him to be the crucial character responsible for the survival of six men in circumstances so precarious that, realistically, the likelihood of survival was virtually nonexistent.

The Murray Levick I have come to know was not so much slow, as Wilfred Bruce and others asserted, as he was preoccupied. He lived in his own mind. How many women did he fuck in his head? How many mental motorbike trips did he take across Europe, Canada, and elsewhere? The Victorian upbringing that had made him great, equally had constrained him. Like his contemporary, Baron Walter Rothschild, Levick worried about airs, about how he would be perceived. It caused him to cover up the bawdy bits about his birds; it caused him to refrain from emotion in his personal life, to resist the urge to leap. And yet, when forced by circumstances to stand up and be counted, it made him more like Amundsen than Scott: calculating, meticulous, detailed. These same qualities also made him ideal as a scientist, an acute observer without the mawkish predispositions that characterized other scientists of his day.

Nothing exemplifies that more than one last piece of Levick's writing that, miraculously, also turns up a century after he wrote it.

In 2012, conservators from the New Zealand Antarctic Program are cleaning out the snow that has accumulated for more than one hundred years beneath Scott's hut at Cape Evans, threatening its foundations. They discover a small brown notebook.

Inside it, there is the same small neat writing that I had first seen in the room of a third-floor apartment in London. It is Levick's, left behind when he "hastened slowly" aboard the *Terra Nova* on January 19, 1913. There is no sex. There are no penguins. It contains, instead, the exposures and details for the processing of each photograph he has taken.

And, in those framed photographs, which lie on the floor all around me as I sit in Levick's chair in his house in Budleigh Salterton, there is everything that you need to know about Levick. He had asked the Terra Nova Expedition's photographer, Herbert Ponting, to teach him photography. Ponting was a photographer of uncompromising excellence but he did not particularly like Levick. Consequently, Levick was left to largely teach himself. His first attempts were appalling: washed out, overexposed, and poorly composed. But Levick had a goal and he was dogged about achieving it. In Carsten Borchgrevink's abandoned and freezing hut at Cape Adare he worked on developing his photographs while George Abbott worked out with a punching bag beside him. He kept notes. In the little brown book. And, eventually, Levick became a highly accomplished photographer, second only to Ponting.

I am a photographer myself. I look critically over the photographs surrounding the chair, left there for me by the current owners of Levick's house. Almost all feature Adelie penguins. They are beautiful. It is as if Levick is most comfortable with facade, like Newcastle upon Tyne never left the boy that was born there. He is happiest showing the world what penguins look like. His photographs have no need for words, no need for codes. People can interpret them as they wish. Only Levick knows the secrets that those photographs contain.

I sit there in Levick's chair, contemplating the significance of those secrets about the penguins' sexual behavior that he had uncovered, and then just as surely covered up again; only for me to walk unknowingly in his footsteps and find the same secrets. Penguins are not the monogamous paragons of virtue we had all thought them to be. But should we think less of them for their sexual excesses, as the Victorian Levick so clearly did?

In many ways, the penguins are as much a product of their environment as they are of their biology. Antarctica is a harsh mistress. She exacts a high price from penguins and men for being with her; for getting things wrong; for being late; for not having enough food, the right feathers, or the right clothes; or, simply, for not being fat enough.

As my research has shown, it is not that evolution finds virtue in homosexuality, divorce, infidelity, rape, or prostitution: these are simply the consequences, the collateral damage, in the competitive race these penguins are in to breed successfully in an environment where the tolerances for success are tiny. Natural selection is all about the winning, not the route taken to get there.

It is us who try to set ourselves apart from nature and pretend we are different. It is us who try to live our lives by a set of religious and social mores. It is us who judge so harshly a man who would use a proven method of dogs and skis to get to the South Pole and with the same breath, create a virtue of failure, of being unprepared in an environment that does not suffer fools either gladly or otherwise.

When I first went to Antarctica in that Summer of '77, in addition to Levick's book and two others about penguins, I took with me my much-loved, much-read copies of the complete works of Henry David Thoreau and Walt Whitman's *Leaves of Grass*. One of the first things I did, when getting to the little green hut at Cape Bird, was to copy out a passage from Thoreau's journals and pin it to the wall above the window in the lab:

Color, which is the poet's wealth, is so expensive that most take
to mere outline or pencil sketches and become men of science.

I wanted it to be a constant reminder to me to not lose sight of the color in that world outside my window. And, I suppose, that's what I wanted my Levick to be: a man of science but with Thoreau's sensibilities. I am sure that is the man Levick himself wanted to be too. The would-be writer, the Second Act, which never quite came. Levick: constrained and quartered by the expectations society had placed upon him; his acceptance of a Victorian moral code that had been foisted on him since birth—the same one that had constrained Scott too, leading to his failure. Whereas men like Nansen, Shackleton, and even Amundsen embraced the color in the world around them irrespective of what society told them they should do.

In that sense, they were probably more like Walt Whitman.

Even in 1977, I had been struck by the superficial resemblance between Whitman's lengthy and famous poem, "Song of Myself," and the ecstatic calls of the male Adelie penguins, which they used to attract mates. When a male Adelie penguin stands on its nest of stones and, bill skyward, calls out to the world, it is a song of "self" no less than Whitman's. But, I see now, from the vantage point of Levick's chair, that the analogy goes far deeper than that. Near the end of his poem, Whitman wrote:

> *Do I contradict myself?*
> *Very well then I contradict myself,*
> *(I am large, I contain multitudes.)*

And by that he meant that he—and by extension, all of us—are a mixture of many contradictions; multitudes, both good and bad.

My investigation of Levick and his observations of penguins, if it has revealed anything at all, it is that Whitman's description of his condition—our condition—should apply as much to penguins as it does people. Our idols are never so virtuous as we make them out to be. The heroes from the Heroic Age could be deceptive, dishonest, and unfaithful. Even a Nobel Prize–winning explorer could have affairs.

I realize, now, that the penguin I fell in love with on my first evening at Cape Bird for being so adorable and fascinating, for carrying himself with such an admirable confidence that he could prosper in such an extreme environment whereas I could not, was undoubtedly also a cruel and ruthless animal. Almost certainly, he would have cheated on his mate if given even half a chance. He may well have fucked chicks or the desiccated corpses of other penguins. He may even have raped.

In weighing these contradictory multitudes in people and penguins, perhaps Whitman's poem does not go far enough? Perhaps Emily and Ernest Shackleton were right all along: it is the Victorian poet Robert Browning, their favorite, to whom we should look for inspiration, if not answers?

The only poem of Browning's with which I am familiar is "Porphyria's Lover," which was ingrained in me at school between lessons about Captain Cook and Joseph Banks. In it, Browning describes a young woman coming into a man's cottage from out of a storm so violent it could have been an Antarctic one. She then proceeds to partially disrobe and begins to seduce him in what is clearly meant as an affront to the Victorian moral codes of the time; those very codes by which Murray Levick was not just brought up, he lived his whole life. But Browning does not stop there: the man strangles the girl with her own hair in order that he might preserve the moment. The poem ends with:

And thus we sit together now,
And all night long we have not stirred,
And yet God has not said a word!

Browning confronts us with beauty, sex, and violence in Victorian society—a concoction that could just as easily describe the Antarctic and the penguins' society that Murray Levick experienced—and he is asking us to judge. Not God. Us. For, despite the admonitions from religious and moral quarters that bind us, the reality is that the sky does not fall in from God's hand. Any censuring comes from us.

Should Sidney Harmer and Levick have felt so compelled to censor the bad bits about the penguins' behavior simply because they feared censure by us? The answer must surely be: no.

In the end, this is what we are left with: that penguins, in their many flawed ways, are really just mirror images of us. As Cherry-Garrard suggested in *The Worst Journey in the World*, the book responsible for my going to Antarctica and thereby studying penguins:

> *All the world loves a penguin: I think it is because in many respects they are like ourselves, and in some respects what we should like to be.*

Certainly, when it comes to sex, as Murray Levick and I have found and the likes of Nansen, Shackleton, and Amundsen have demonstrated, they are even more like us than we could have ever possibly imagined.

And yet God has not said a word!

FURTHER READING: KEY REFERENCES

Amundsen, Roald. *The North West Passage: Being the Record of a Voyage of Exploration of the ship "Gjoa" 1903–1907*. London: Archibald Constable and Co., Ltd, 1908.

Amundsen, Roald. *The South Pole, Volumes 1 and 2: An Account of the Norwegian Antarctic Expedition in the "Fram," 1910–1912. Translated from the Norwegian by A. G. Chater*. London: John Murray, 1912.

Beattie, Owen and John Geiger. *Frozen in Time: The Fate of the Franklin Expedition*. London: Bloomsbury Publishing, 1987.

Bernacchi, Louis. *To the South Polar Regions: Expedition of 1898–1900*. London: Hurst and Blackett, Ltd., 1901.

Bomann-Larsen, Tor. *Roald Amundsen*. Stroud, UK: The History Press, 2011.

Borchgrevink, C.E. *First on the Antarctic Continent: Being an Account of the British Antarctic Expedition 1898–1900*. London: George Newnes Ltd, 1901.

Bown, Stephen. *The Last Viking: The Life of Roald Amundsen, Conqueror of the South Pole*. Philadelphia: Da Capo Press, 2012.

Byrd, Richard Evelyn. *Discovery: The Story of the Second Byrd Antarctic Expedition*. New York: G.P. Putnam's Sons, 1935.

Campbell, Victor. *The Wicked Mate: The Antarctic Diary of Victor Campbell, an Account of the Northern Party on Captain Scott's Last Expedition from the Original Manuscript in the Queen Elizabeth II Library, Memorial University of Newfoundland*. Alburgh, UK: Bluntisham Books/Erskine Press, 1988.

Cherry-Garrard, Apsley. *The Worst Journey in the World: Antarctic 1910–1913*. London: Chatto & Windus, 1922.

Cook, James. *A Voyage Towards the South Pole and Round the World: Performed in His Majesty's Ships the Resolution and Adventure, in the Years 1772, 3, 4 and 5.* London: W. Stahan and T. Caldwell, 1777.

Davis, John King. *High Latitude.* Melbourne: Melbourne University Press, 1962.

Davis, Lloyd Spencer. *Professor Penguin: Discovery and Adventure with Penguins.* Auckland, New Zealand: Random House, 2014.

Day, David. *Antarctica: a biography.* North Sydney: Knopf, 2012.

Day, David. *Flaws in the Ice: In search of Douglas Mawson.* Melbourne: Scribe, 2013.

Dickason, Harry. *Penguins and Primus: An Account of the Northern Expedition June 1910 – February 1913.* Perth: Australian Capital Equity/Freemantle Press, 2013.

Fitzsimons, Peter. *Mawson and the Ice Men of the Heroic Age: Scott, Shackleton and Amundsen.* Australia: William Heinemann, 2011.

Gran, Tryggve. *The Norwegian with Scott: Tryggve Gran's Antarctic Diary 1910– 1913*, edited by Geoffrey Hattersley-Smith, translated by Ellen Johanne McGhie. London: HMSO Books, 1984.

Guly, Henry R. "George Murray Levick (1876–1956), Antarctic explorer," *Journal of Medical Biography* 24, no. 1 (2014): 4–10. doi. org/10.1177/0967772014533051

Herbert, Kari. *Heart of the Hero: The Remarkable Women who Inspired the Great Polar Explorers.* Glasgow: Saraband, 2013.

Hooper, Meredith. *The Longest Winter: Scott's Other Heroes.* London: John Murray, 2010.

Huntford, Roland. *Scott and Amundsen: The Last Place on Earth.* London: Little, Brown Book Group, 1979.

Huntford, Roland. *Shackleton.* London: Little, Brown Book Group, 1985.

Hurley, Captain Frank. *Argonauts of the South: Being a Narrative of Voyagings and Polar Seas and Adventures in the Antarctic with Sir Douglas Mawson and Sir Ernest Shackleton.* New York: G.P. Putnam's Sons, The Knickerbocker Press, 1925.

Huxley, Leonard, ed. *Scott's Last Expedition: Volume II, Being the Reports of the Journeys and the Scientific Work Undertaken by Dr. E. A. Wilson and the surviving members of the Expedition.* London: Macmillan and Co., Ltd, 1913.

Kennet, Lady Kathleen. *Self-Portrait of an Artist: From the Diaries and Memoirs of Lady Kennet, Kathleen, Lady Scott.* London: John Murray, 1949.

Kløver, Geir O. *Lessons from the Arctic: how Roald Amundsen won the race to the South Pole.* Oslo: The Fram Museum, 2017.

Lambert, Katherine. *'Hell with a Capital H': An Epic Story of Antarctic Survival.* London: Pimlico, 2002.

Levick, G. Murray. *Antarctic Penguins: a study of their social habits.* London: William Heinemann, 1914.

Levick, George Murray. *A Gun for a Fountain Pen: Antarctic Journal November 1910 – January 1912.* Perth: Freemantle Press, 2013.

Mawson, Sir Douglas. *The Home of the Blizzard: being the story of the Australasian Antarctic expedition, 1911–1914.* London: Ballantyne Press, 1915.

Murray, James and George Marston. *Antarctic Days: Sketches of the homely side of Polar life by two of Shackleton's men.* London: Andrew Melrose, 1913.

Nansen, Fridtjof. *Farthest North: Being the Record of a Voyage of Exploration of the Ship "Fram" 1893-96 and of a Fifteen Months' Sleigh Journey by Dr. Nansen and Lieut. Johansen.* London: Harper and Brothers Publishers, 1897.

Ponting, Herbert. *The Great White South, or, With Scott in the Antarctic : being an account of experiences with Captain Scott's South Pole expedition and of the nature life of the Antarctic.* London: Duckworth & Co., 1923.

Ponting, Herbert. *With Scott to the Pole: The Terra Nova Expedition 1910 – 1913, The Photographs of Herbert Ponting.* Crows Nest, NSW, Australia: Allen & Unwin, 2004.

Preston, Diana. *A First Rate Tragedy: Robert Falcon Scott and the Race to the South Pole.* Boston: Houghton Mifflin Co., 1998.

Priestley, Raymond E. *Antarctic Adventure: Scott's Northern Party.* London: T. Fisher Unwin, 1914.

Riffenburgh, Beau. *Shackleton's Forgotten Expedition: The Voyage of the Nimrod.* New York: Bloomsbury, 2005.

Riffenburgh, Beau and Geir Kløver. *Carsten Borchgrevink and the Southern Cross Expedition, 1898–1900.* Oslo: The Fram Museum, 2017.

Riffenburgh, Beau and Geir O. Kløver. *Eivind Astrup: The Norwegian Ski and Sledge Expert with Peary.* Oslo: The Fram Museum, 2000.

Russell, Douglas G.D., William J.L. Sladen, and David G. Ainley. "Dr. George Murray Levick (1876–1956): unpublished notes on the sexual habits of the Adélie penguin." *Polar Record* 48, no. 247 (2012): 387–393. doi:10.1017/S0032247412000216.

Ruzesky, Jay. *In Antarctica: An Amundsen Pilgrimage.* Gibsons, Canada: Nightwood Editions, 2013.

Scott, Captain Robert F. *The Voyage of the 'Discovery'.* London: Smith, Elder, & Co., 1905.

Scott, Captain R. F. *Scott's Last Expedition: Volume I, Being the Journals of Captain R.F. Scott, R.N., C.V.O.* London: Macmillan and Co., Ltd, 1913.

Seaver, George. *Edward Wilson of the Antarctic: Naturalist and Friend.* London: John Murray, 1933.

Shackleton, E. H. *The Heart of the Antarctic: being the story of the British Antarctic Expedition 1907–1909.* London: William Heinemann, 1909.

Shackleton, Sir Ernest. *South: The Story of Shackleton's 1914-1917 Expedition.* London: William Heinemann, 1919

Smith, Michael. *I Am Just Going Outside: Captain Oates – Antarctic Tragedy.* Staplehurst, UK: Spellmount Ltd., 2008.

Smith, Michael. *Shackleton: By Endurance We Conquer*. Cork, Ireland: The Collins Press, 2014.

Utne, Eric, ed. *Brenda My Darling: The Love Letters of Fridtjof Nansen to Brenda Ueland*. Minneapolis: Utne Institute, 2011.

Watson, Paul. *Ice Ghosts: The Epic Hunt for the Lost Franklin Expedition*. New York: W. W. Norton & Co., Ltd, 2017.

Wheeler, Sara. *Cherry: A Life of Apsley Cherry-Garrard*. London: Jonathan Cape, 2001.

Wilson, Edward. *Diary of the Terra Nova Expedition to the Antarctic 1910–1912*. London: Blandford Press, 1972.

Young, Louisa. *A Great Task of Happiness: The Life of Kathleen Scott*. London: Macmillan, 1995.

ACKNOWLEDGMENTS

I am indebted to Douglas Russell for his generosity and kindness in sharing his discovery of Murray Levick's unpublished paper with me and for providing me with the key ingredient needed for any story: access. Douglas is also an expert on Apsley Cherry-Garrard and his insights have been a considerable help.

I am equally indebted to Richard Kossow for his trust and support of me: without it, this book simply would not exist. His deep love and respect for the Heroic Age of Antarctic exploration has ensured that many treasures, including those of Murray Levick, have been preserved for the benefit of all of us.

I thank the librarians at the Scott Polar Research Institute in Cambridge, the Queen Elizabeth II Library at the Memorial University of Newfoundland, The Keep of the East Sussex Records Office, the Alexander Turnbull Library in Wellington and the Port Chalmers Library for helping me with access to archived materials. Justin Warwick at the British Exploring Society kindly offered me unfettered access to their archives. The Antarctic Heritage Trust in Christchurch provided me with details concerning the discovery of Levick's photography notebook under the hut at Cape Evans. Dr Geir Kløver of the Fram Museum in Oslo was exceptionally generous both in terms of his time and in sharing his extensive knowledge of polar exploration. Meredith Hooper, an expert on the Northern Party, helped me figure out some crucial details.

ACKNOWLEDGMENTS

Roger and Judith Kingwill, Mike and Margaret Wilson, and Michael Downes were instrumental in making my stay in Budleigh Salterton a pleasant and productive one, as well as providing key details about Murray Levick and his family. My good friends, Andy and Sarah Wroot, kindly put up with me imposing upon them while I was in the UK conducting research.

Rodney Russ of Heritage Expeditions generously took me to the Ross Sea area on the *Shokalskiy* in order that I might follow in the "footsteps" of Murray Levick. I am especially thankful to the Stuart Residence Halls Council at the University of Otago for giving me the means and support necessary for researching this book. I am also indebted to my many students and postdoctoral fellows who have contributed to my research over four decades: thank you all!

Russell Galen, my literary agent and best critic, has been relentless in pushing me in ways that have improved my writing and, in particular, this story, and for that I am forever grateful. Jessica Case, my editor at Pegasus Books, has been a wonderful champion for this book and much of its "feel" owes a lot to her touch. Drew Wheeler provided expert input as a copy editor.

My partner, Wiebke Finkler, has provided me with the most essential and most appreciated ingredient for this book: her love and support, which has allowed me to take the endless hours from the family that it has, inevitably, exacted. Finally, I thank my children—Daniel, Kelsey, and Eligh—for their love and support: the best legacy a guy could ever have.

ENDNOTES

PROLOGUE

p. ix "'1st Penguin arrives Oct. 13th,' which he underlines with a sweep of the blue-black ink."; George Murray Levick, *Zoological Notes Cape Adare Vol. 1.* (Unpublished).

p. x "*We shall stick it out to the end, but we are getting weaker, of course, and the end cannot be far*"; Captain R. F. Scott, *Scott's Last Expedition: Volume I, Being the Journals of Captain R. F. Scott, R.N., C.V.O.* (London: Macmillan and Co., Ltd., 1913).

p. xi "He picks up his own pencil and writes, 'Blowing hard all day.'"; George Murray Levick, diaries. (Scott Polar Research Institute, Unpublished).

PART ONE: THE LURE OF ANTARCTICA

ONE: VICTORIAN VALUES

p. 8 "*Here on one occasion I saw what I took to be a cock copulating with a hen.*"; Douglas G. D. Russell, William J. L. Sladen, and David G. Ainley. "Dr. George Murray Levick (1876–1956): unpublished notes on the sexual habits of the Adélie penguin," *Polar Record* 48, no. 247 (2012): 387–393. doi:10.1017/S0032247412000216.

TWO: TERRA AUSTRALIS

p. 24 "*Harmer writes to Ogilvie-Grant and says we'll have it cut out, we're not going to include it in the published version.*"; Douglas Russell, recorded interview with author, Natural History Museum, Tring, July 22, 2013.

THREE: THE THREE NORWEGIANS

p. 32 "Roald Amundsen is eight or nine when the story of Sir John Franklin's lost expedition 'captivated his imagination.'"; Roald Amundsen, *The North West Passage: Being the Record of a Voyage of Exploration of the ship "Gjoa" 1903–1907* (London: Archibald Constable and Co., Ltd., 1908).

p. 33 "*That day I wandered with throbbing pulses amid the bunting and the cheers, and all my boyhood's dreams reawoke to tempestuous life.*"; Ibid.

p. 36 "*here the unbound forces of the Antarctic Circle do not display the whole severity of their powers.*"; C. E. Borchgrevink, *First on the Antarctic Continent: Being an*

Account of the British Antarctic Expedition 1898–1900 (London: George Newnes Ltd., 1901).

p. 39   "In it, he ascribes to Amundsen the comment, 'Yes, sir, I love it' when referring to the absence of women."; Roland Huntford, *Scott and Amundsen: The Last Place on Earth* (London: Little, Brown Book Group, 1979).

p. 40   "*I stood in Nansen's villa at Liysaker and knocked on the door of his study.*"; Amundsen, *The North West Passage*.

PART TWO: ALL ROADS LEAD TO CAPE ADARE

FOUR: FIRST OBSERVATIONS

p. 47   "*It seemed, at a distance, so small and inhospitable that some of my staff felt constrained to remark at first sight of the place*"; Borchgrevink, *First on the Antarctic Continent*.

p. 49   "they have, as Bernacchi puts it, 'only succeeded in excavating to a depth of about 4 inches.'"; Louis Bernacchi, *To the South Polar Regions: Expedition of 1898–1900* (London: Hurst and Blackett, Ltd., 1901).

p. 49   "*It was one of the most bleak and ungenial days imaginable . . . We were not sorry to leave that gelid desolate spot*"; Ibid.

p. 50   "a large iceberg calves off from the nearby glacier and 'some thousands of tons of ice fell into the sea with a terrific and reverberating roar,'"; Ibid.

p. 51   "reaching 78°50'S, 'the farthest south ever reached by man' as he will boast proudly afterward."; Borchgrevink, *First on the Antarctic Continent*.

p. 51   "*We all watched the life of the penguins with the utmost interest, and I believe and hope that some of us learnt something from their habits and characteristics*"; Ibid.

p. 55   "many of their number form 'their nests on the steep hillsides, even to a height of 1,000 feet"; Scott, *The Voyage of the 'Discovery'*.

p. 55   "He is equally impressed with the first human nest, noting that Borchgrevink's hut 'is in very good condition'"; Ibid.

p. 55   "*There is always something sad in contemplating the deserted dwellings of mankind, under whatever conditions the inhabitants may have left.*"; Ibid.

p. 55   "*There were literally millions of them. It simply stunk like hell, and the noise was deafening.*"; Beau Riffenburgh, *Shackleton's Forgotten Expedition: The Voyage of the Nimrod* (New York: Bloomsbury, 2005).

p. 56   "*The honour of being the first aeronaut to make an ascent in the Antarctic regions, perhaps somewhat selfishly, I chose for myself.*"; Scott, *The Voyage of the 'Discovery'*.

p. 56   "The Captain, knowing nothing whatever about the business, insisted on going up first and through no fault of his own came back safely."; Huntford, *Scott and Amundsen*.

p. 58   "*it left in each one of our small party an unconquerable aversion to the employment of dogs in this ruthless fashion*"; Scott, *The Voyage of the 'Discovery'*.

FIVE: BOYHOOD DREAMS

p. 66   "*we find the Emperor penguin hatching out its chicks in the coldest month of the whole Antarctic year*"; Edward A. Wilson. *Appendix II. On the Whales, Seals, and Birds of*

Ross Sea and South Victoria Land, in *The Voyage of the 'Discovery'* (London: Smith, Elder, & Co., 1905), 352–374.

p. 72   "and about how superior dogs were to what he called the 'futile toil' of man-hauling."; Huntford, *Scott and Amundsen*.

p. 72   *"We ourselves tried some substantial steaks and found the meat excellent."* Ibid.

p. 72   "Amundsen hoists the *Gjoa's* flag and they 'went by the grave in solemn silence.'"; Amundsen, *The North West Passage*.

p. 72   *"The North-West Passage had been accomplished—my dream from childhood."*; Ibid.

SIX: LOST OPPORTUNITIES

p. 76   "In 1907, Ernest Shackleton is determined to prove himself 'a better man than Scott.'"; Peter Fitzsimons, *Mawson and the Ice Men of the Heroic Age: Scott, Shackleton and Amundsen* (Australia: William Heinemann, 2011).

p. 76   "McMurdo Sound and the Ross Island area of Antarctica, which he regards as his 'own field,' as he puts it."; Robert Falcon Scott, letter to Ernest Shackleton, March 18, 1907, Scott Polar Research Institute, MS1456/23.

p. 80   *"To the biologist, no more uninviting desert is imaginable than Cape Royds"*; James Murray, *Appendix One. Biology: Notes by James Murray, Biologist of the Expedition*, in *The Heart of the Antarctic: Being the Story of the British Antarctic Expedition 1907–1909* (London: William Heinemann, 1909).

p. 80   *"There is endless interest in watching them, the dignified Emperor, dignified notwithstanding his clumsy waddle"*; Ibid.

p. 85   *"the Adelie appears to be entirely moral in his domestic arrangements"*; Ibid.

SEVEN: COURTSHIP

p. 90   *"Throw up your cap & shout & sing triumphantly, meseems we are in a fair way to achieve my end."*; Huntford, *Scott and Amundsen*.

p. 91   "As he put it in a letter to her, it is 'a man's way to want a woman altogether to himself.'"; Michael Smith, *Shackleton: By Endurance We Conquer* (Cork, Ireland: The Collins Press, 2014).

p. 91   *"Though the grip of the frost may be cruel and relentless its icy hold"*; Eleanor Harding, *"Shackleton's Secret Lover: Polar Explorer Was So Smitten He Named a Mountain after Her,"* *Daily Mail*, November 12, 2011.

p. 92   "It is mounted in silver metal with a small plaque that reads 'Summit Mount Hope'"; Ibid.

EIGHT: DECEPTION

p. 101   "to 'reach the South Pole and secure for the British Empire the honor of that achievement.'"; Diana Preston, *A First Rate Tragedy: Robert Falcon Scott and the Race to the South Pole* (Boston: Houghton Mifflin Co., 1998).

p. 102   "its primary focus is 'oceanographic investigation' rather than being first to the North Pole"; *Los Angeles Times*, September 3, 1909, 14.

p. 102   "'a mystic look softened his eyes, the look of a man who saw a vision.'"; Hugh Robert Mill, *The Life of Sir Ernest Shackleton* (London: William Heinemann Ltd., 1923).

p. 104   "a loose page has been inserted merely saying, 'The Pole at last!!!'"; Boyce Rensberger, "National Geographic Reverses, Agrees Adm. Peary Missed North Pole," *Washington Post*, September 18, 1988.

p. 106   *"I am preparing a purely Scientific Expedition to operate along the coast of Antarctica commencing in 1911"*; Ernest Shackleton, letter to Robert Falcon Scott, February 21, 1910, Scott Polar Research Institute, MS367/17/2.

p. 109   *"We left Cardiff weather fine and calm."*; Victor Campbell, *The Wicked Mate: The Antarctic Diary of Victor Campbell, an Account of the Northern Party on Captain Scott's Last Expedition from the Original Manuscript in the Queen Elizabeth II Library, Memorial University of Newfoundland* (Alburgh, UK: Bluntisham Books/ Erskine Press, 1988).

p. 110   "He would call it 'the bitterest moment in my life.'"; Huntford, *Scott and Amundsen.*

p. 111   *"I beg your forgiveness for what I have done"*; Ibid.

NINE: THE EASTERN PARTY

p. 113   "He seems quite incapable of learning anything fresh"; Katherine Lambert, *'Hell with a Capital H': An Epic Story of Antarctic Survival* (London: Pimlico, 2002).

p. 115   *"I hope it will never fall to my lot to have more than one wife at a time to look after."*; Edward Wilson, *Diary of the Terra Nova Expedition to the Antarctic 1910–1912* (London: Blandford Press, 1972).

p. 115   "BEG LEAVE TO INFORM YOU FRAM PROCEEDING ANTARCTIC AMUNDSEN"; Tryggve Gran, *The Norwegian with Scott: Tryggve Gran's Antarctic Diary 1910–1913*, edited by Geoffrey Hattersley-Smith, translated by Ellen Johanne McGhie (London: HMSO Books, 1984).

p. 115   "Nansen's reply is even more concise than Amundsen's cable had been: UNKNOWN."; Ibid.

p. 116   *"I think most of us feel regrets a (sic) leaving New Zealand, as we have all made friends"*; George Murray Levick, *A Gun for a Fountain Pen: Antarctic Journal November 1910–January 1912* (Perth: Freemantle Press, 2013).

p. 117   *"there was more blood and hair flying about the hotel than you would see in a Chicago slaughter house in a month"*; Titus Oates, letter to his mother, November 23, 1910, Scott Polar Research Institute, MS1016/337/1.

p. 119   "Campbell notes that, 'We must hope for fine passage,' but that is not to be."; Campbell, *The Wicked Mate.*

p. 119   "which as Campbell nonchalantly observes is, "very slow work as the men were constantly being washed off their legs"; Ibid.

p. 120   "More tender than beef steak and quite as good to eat"; Levick, *A Gun for a Fountain Pen.*

p. 120   *"a really first class bird—rather like blackcock to taste, but a good deal better"*; Ibid.

p. 120   "the men sang, 'in a horrible discordant manner to Adelie penguins that had gathered about the stationary ship"; Ibid.

p. 120   "the penguins stood around, 'cawing and bowing their appreciation.'"; Ibid.

p. 120   "Herbert Ponting, the expedition's photographer, or 'camera artist'"; Lambert, *'Hell with a Capital H.'*

p. 121   "Levick admiringly notes, 'To his great credit he saved his camera and tripod.'";
Levick, *A Gun for a Fountain Pen.*

p. 121   *"I find I can't get any information out of Ponting—He won't give anything away."*; Ibid.

p. 121   "Always I have had the feeling that Cape Royds has been permanently
vulgarized."; Meredith Hooper, *The Longest Winter: Scott's Other Heroes* (London:
John Murray, 2010).

p. 122   "'There is no trail of Shackleton there,' he says to her"; Ibid.

p. 122   *"In the middle of the hut was a long table with the remains of their last meal"*; Levick,
A Gun for a Fountain Pen.

p. 122   "Priestley, who found the experience of going back to the hut where he had lived
'very eerie.'"; Lambert, *'Hell with a Capital H.'*

p. 122   *"I expect to see people come in through the door after a walk over the surrounding
hills"*; Ibid.

p. 122   "adult birds 'bringing in food for their little downy youngsters'; Levick, *A Gun for
a Fountain Pen.*

p. 122   "been so well described by Wilson in his 'Discovery' reports that it is no good
repeating them here . . .'"; Ibid.

p. 124   "At least Priestley is glad to have 'set the matter at rest finally.'"; Hooper, *The
Longest Winter.*

p. 124   "after he is woken and rushes on deck with the rest, 'None of us needed to be told
that it was the "Fram."'"; Levick, *A Gun for a Fountain Pen.*

p. 125   "all of them now waving their arms like 'incurable lunatics.'"; Roald Amundsen,
*The South Pole, Volumes 1 and 2: An Account of the Norwegian Antarctic Expedition
in the "Fram," 1910–1912. Translated from the Norwegian by A. G. Chater*
(London: John Murray, 1912).

p. 125   *"We had talked of the possibility of meeting the Terra Nova . . . but it was a great
surprise all the same."*; Ibid.

p. 125   "He looks older than Campbell expected, a 'fine looking man' with 'hair nearly
white.'"; Campbell, *The Wicked Mate.*

p. 125   *"I think that no incident was so suggestive of the possibilities latent in these teams
as the arrival of Amundsen at the side of the Terra Nova."*; Raymond E. Priestley,
Antarctic Adventure: Scott's Northern Party (London: T. Fisher Unwin, 1914).

p. 125   "The principal trump-card of the Norwegians was undoubtedly their splendid
dogs."; Ibid.

p. 126   *"We found them all men of the of the very best type, and got on very well."*; Levick, *A
Gun for a Fountain Pen.*

p. 126   "others argue against it on the grounds that, 'the feelings between the two
expeditions must be strained'"; Ibid.

p. 126   "In summing it up in his diary, he writes, 'This has been a wonderful day.'"; Ibid.

PART THREE: CAPE ADARE

TEN: THE NORTHERN PARTY

p. 132   *"On this little patch of peninsular (sic), about a triangular mile in extent, we are
absolute prisoners"*; Levick, *A Gun for a Fountain Pen.*

p. 132   "Campbell wrote wryly in his log, 'No sign of a possible landing anywhere.'";
Campbell, *The Wicked Mate.*

p. 132   "Borchgrevink's hut is still standing there 'in good preservation'"; Levick, *A Gun for a Fountain Pen*.

p. 133   "the sixteen carcasses of mutton they have unloaded from the ship are 'covered with green mould.'"; Campbell, *The Wicked Mate*.

p. 133   *"I am sorry to say that a great many visitors we knock on the head and put in the larder"*; Ibid.

p. 133   "shown up at the hut in his fancy new suit of feathers only to be 'taken for a stranger and killed' by the men"; Priestley, *Antarctic Adventure*.

p. 133   "the men move into their new quarters with much fanfare that includes 'a great house warming"; Campbell, *The Wicked Mate*.

p. 134   *"dear little things, and I hate having to kill them"*; Levick, *A Gun for a Fountain Pen*.

p. 135   "The blizzards are only 'a pleasant rest' for the dogs"; Scott, *Scott's Last Expedition*.

p. 135   "They are curled snugly under the snow and at meal times issue from steaming warm holes."; Ibid.

p. 135   "Meanwhile, the ponies are suffering: 'so frozen they can hardly eat,' observes Gran"; Gran, *The Norwegian with Scott*.

p. 135   "But Scott will not tolerate perpetuating 'this cruelty to animals,' as he calls it"; Huntford, *Scott and Amundsen*.

p. 135   "Oates says, 'I'm afraid you will regret it, Sir.'"; Ibid.

p. 136   "Bowers may well lament that he is left 'carrying a blood stained pick-axe instead of leading the pony'"; Apsley Cherry-Garrard, *The Worst Journey in the World: Antarctic 1910–1913* (London: Chatto & Windus, 1922).

p. 136   "to to protect the ponies from the cold, 'It makes a late start necessary for next year.'"; Scott, *Scott's Last Expedition*.

p. 140   *"It surprises me very much to hear that Captain Scott has landed a party at Cape Adare"*; *Daily Mail*, March 28, 1911, 9.

p. 140   *"For an hour or so we were furiously angry, and were possessed with an insane sense that we must go straight to the Bay of Whales"*; Cherry-Garrard, *The Worst Journey in the World*.

ELEVEN: THE WORST JOURNEY

p. 142   "they manage to celebrate in style with champagne, brandy, cigars, and 'an extended sing-song'"; Levick, *A Gun for a Fountain Pen*.

p. 142   *"Inside the hut are orgies. We are very merry—and indeed why not?"*; Cherry-Garrard, *The Worst Journey in the World*.

p. 143   *"It took two men to get one man into his harness, and was all they could do, for the canvas was frozen"*; Ibid.

p. 144   *"The horror of the nineteen days it takes us to travel from Cape Evans to Cape Crozier"*; Ibid.

p. 144   "writes Cherry-Garrard, 'but when your body chatters you may call yourself cold.'"; Ibid.

p. 146   "they try again, this time near the middle of the day when the darkness is not so 'pitchy black.'"; Ibid.

p. 146   *"After indescribable effort and hardship we were witnessing a marvel of the natural world"*; Ibid.

p. 147   "*we on this journey were already beginning to think of death as a friend*"; Ibid.

p. 147   "*it was blowing as though the world was having a fit of hysterics*"; Ibid.

p. 149   "*In these poor birds the maternal side seems to have necessarily swamped the other functions of life*"; Ibid.

p. 150   "*I might have speculated on my chances of going to Heaven; but candidly I did not care*"; Ibid.

TWELVE: THE RELUCTANT PENGUIN BIOLOGIST

p. 154   "As he writes, in a way that appears starker when written in his cursive script in the blue-black ink, 'I killed them all'"; Levick, *Zoological Notes.*

p. 154   "and its stomach is full of 'surprisingly large' basalt stones and nothing else"; Ibid.

p. 154   "Dickason describes it as, 'the hardest blow we have had'"; Harry Dickason, *Penguins and Primus: An Account of the Northern Expedition June 1910–February 1913* (Perth: Australian Capital Equity/Freemantle Press, 2013).

p. 155   "*we gazed seaward this morning and realized the astounding fact that the sea ice beyond the bay*"; Priestley, *Antarctic Adventure.*

p. 156   "attempting to get to Cape Wood, they 'would have certainly been dead men'"; Levick, *A Gun for a Fountain Pen.*

p. 156   "We cannot leave as long as the temperature keeps so low . . . it will be terrible for the dogs"; Huntford, *Scott and Amundsen.*

p. 157   "*I don't call it an expedition. It's panic.*"; Ibid.

p. 158   "*I find it most correct with the good of the expedition in view—to dismiss you from the journey to the South Pole* "; Ibid.

p. 159   "Levick describes it as 'the most trying night I have ever spent'"; Levick, *A Gun for a Fountain Pen.*

p. 159   "'A heartless business,' as Levick calls it, but he justifies it as a kind of necessary cruelty."; Ibid.

p. 160   "*A dead black throated penguin lay with a rope tied round one of its legs*"; Levick, *Zoological Notes.*

p. 160   "*the Finn Savio conceived rather a good idea of amusing himself*"; Borchgrevink, *First on the Antarctic Continent.*

p. 161   "*Levick had started on a series of systematic notes, which are probably the most thorough that have ever been made*; Priestley, *Antarctic Adventure.*

p. 161   "*Have been reading up all I can find about penguins*"; Lambert, '*Hell with a Capital H.*'

p. 162   "*Never write down anything as a fact unless you are absolutely certain.*"; Priestley, *Antarctic Adventure.*

p. 162   "*he thinks much less of the men who would perpetrate what he calls, the "scene of tragedy I saw*"; Levick, *Zoological Notes.*

p. 163   "each with about fifty couples, for special observation"; Ibid.

p. 163   "intention is to remove all the eggs from this group as they are layed (sic)"; Ibid.

p. 163   "A couple of minutes after I removed the eggs, the owners seemed to have forgotten the incident entirely"; Ibid.

p. 164   "He describes Levick as 'the slowest man I've ever met'"; Hooper, *The Longest Winter.*

THIRTEEN: THE RACE BEGINS

p. 166   "It is 'a brilliant test,' he notes, of their meticulous precautions and preparations.";
Huntford, *Scott and Amundsen.*

p. 166   "they are, as Amundsen records on October 24, 1911, 'enjoying life.'"; Ibid.

p. 167   *"A more unpromising lot of ponies to start a journey such as ours it would be almost
impossible to conceive";* Titus Oates, Note, October 31, 1911, Scott Polar Research
Institute, MS1317/2.

p. 167   *"Scott realizes now what awful cripples our ponies are and carries a face like a
tired old sea boat in consequence";* Michael Smith, *I Am Just Going Outside:
Captain Oates–Antarctic Tragedy* (Staplehurst, UK: Spellmount Ltd., 2008);
but see Huntford, *Scott and Amundsen,* who quotes "seaboot" rather than "sea
boat."

p. 168   "even Scott thinks 'Amundsen with his dogs may be doing much better.'";
Huntford, *Scott and Amundsen.*

p. 169   "the penguins throughout the colony have become 'noticeably subdued' with very
little activity at all"; Levick, *Zoological Notes.*

p. 169   "the penguins resume their frantic 'love making, fighting, and building' of their
nests."; Ibid.

p. 169   *"the roar of battles & thuds of blows can be heard over the entire rookery";* Ibid.

p. 169   *"I conclude when I see two birds fight with flippers alone, that they are cocks.";* Ibid.

p. 169   "he sees a penguin with its eye 'put out' by another's beak"; Ibid.

p. 169   *" it was not unusual to see a strange cock paying court to a mated hen in the absence of
her husband";* G. Murray Levick, *Antarctic Penguins: A Study of Their Social Habits*
(London: William Heinemann, 1914).

p. 170   *"the mated cock has suddenly turned up and fought the interloper";* Levick, *Zoological
Notes.*

p. 170   "male penguins newly arrived at the colony, as evidenced by their 'spotlessly clean'
white breasts"; Ibid.

p. 170   "marking her in Levick's estimation as 'unquestionably an old arrival and a bride
long past her virginity.'"; Ibid.

p. 171   "a neighbor 'put out its beak and stole one of the pieces!'"; Ibid.

p. 171   "he laments, 'Unfortunately I am going away sledging for four days . . .'"; Ibid.

p. 172   "the penguin is trying to sit on eggs 'amidst a slush of melting snow, so that the
eggs were nearly floating in water.'"; Ibid.

p. 172   "the most striking fact about this rookery seemed to me to be the absence of open
water for many miles"; Ibid.

p. 172   "by November 20 he laments that 'my photography (chiefly in work with
Priestley) is taking a great amount of time'"; Ibid.

p. 172   "Mated couples appear to fast absolutely until the first egg is laid, after which they
go off to feed by turns."; Ibid.

p. 173   "the couples took turn and turn about on the nest, one remaining to guard and
incubate while the other went off to the water."; Levick, *Antarctic Penguins.*

p. 174   *"they were sociable animals, glad to meet one another, and, like many men, pleased
with the excuse to forget for a while their duties at home";* Ibid.

p. 174   "Levick records that the skuas 'are stealing a large number of eggs.'"; Levick,
Zoological Notes.

p. 174   "he notes that 'a large number of nests in the rookery are now to be seen deserted.'"; Ibid.

p. 174   "eggs in some nests 'may have first been filched by skuas, and the nest then deserted'"; Ibid.

p. 174   "he emphasizes that 'the number of deserted nests is now very great indeed.'"; Ibid.

p. 174   "he states that 'The number of deserted nests continues to increase.'"; Ibid.

p. 174   "as many are to be found on the ground, frozen, which have not yet been eaten by skuas"; Ibid.

p. 175   *"I daresay the cocks are the greater offenders in this respect . . ."*; Ibid.

FOURTEEN: COMPETITION

p. 177   "Amundsen chooses an especially daunting one: a steep, wide glacier marked by 'crevasses and chasms,'"; Huntford, *Scott and Amundsen*.

p. 178   "what Amundsen describes as 'pit after pit, crevasse after crevasse, and huge ice blocks scattered helter skelter.'"; Ibid.

p. 178   *"Glittering white, shining blue, raven black . . . the land looks like a fairytale."*; Ibid.

p. 178   "there was depression and sadness in the air—we had grown so fond of our dogs"; Amundsen, *The South Pole*.

p. 182   *"This afternoon I saw a most extraordinary site (sic)."*; Russell et al., Dr. George Murray Levick (1876–1956).

p. 182   *"On returning to the hut I told Browning, hardly expecting to be believed"*; Ibid.

p. 183   "there must be a certain number of both cocks and hens wandering about who have been left out in the race for partners"; Levick, *Zoological Notes*.

p. 184   "he is so prepared to sacrifice the ponies, which Cherry-Garrard describes as 'a horrid business.'"; Cherry-Garrard, *The Worst Journey in the World*.

p. 185   "Amundsen has declared it a rest day 'to prepare for the final onslaught.'"; Huntford, *Scott and Amundsen*.

p. 185   "'Sledges and ski glide easily and pleasantly,' according to Amundsen"; Ibid.

FIFTEEN: TIMING

p. 188   "'owing to the wind the old birds are sitting very closely and there are probably many hatched already.'"; Levick, *Zoological Notes*.

p. 188   *"Whilst the chicks are small the two parents manage to keep them fed without much difficulty"*; Levick, *Antarctic Penguins*.

p. 188   *"To see an Adélie chick of a fortnight's growth trying to get itself covered by its mother is a most ludicrous sight."*; Ibid.

p. 190   *"I think somehow we are the first to see this curious sight."*; Huntford, *Scott and Amundsen*.

p. 190   *"I have never known any man to be placed in such a diametrically opposite position to the goal of his desires"*; Amundsen, *The South Pole*.

p. 191   *"Ski are the thing, and here are my tiresome fellow-countrymen too prejudiced to have prepared themselves for the event."*; Scott, *Scott's Last Expedition*.

p. 192   "'the perfect mass of crevasses into which we all continually fall; mostly one foot, but often two, and occasionally we went down altogether.'"; Cherry-Garrard, *The Worst Journey in the World*.

p. 192   "Scott tells Atkinson 'to bring the dog-teams out to meet the Polar Party'"; Ibid.

p. 192   *"The final advance to the Pole was, according to plan, to have been made by four men."*; Ibid.

p. 193   "'We started more than half an hour later on each march and caught the others easy. It's been a plod for the foot people and pretty easy going for us.'"; Scott, *Scott's Last Expedition.*

p. 193   "At present everything seems to be going with extraordinary smoothness."; Ibid.

p. 194   "He hoisted 'the flag to signal to the hut.'"; Levick, *A Gun for a Fountain Pen.*

p. 194   *"The penguin chicks are able to walk now and huddle together in batches."*; Campbell, *The Wicked Mate.*

PART FOUR: AFTER CAPE ADARE

SIXTEEN: HOOLIGANS

p. 204   *"The cock did not seize the hen with his beak, by the feathers on the back of her head as chickens do."*; Levick, *Zoological Notes.*

p. 204   "There is also an added note, written in a different light blue ink that says, "More notes on this later."; Ibid.

p. 204   "At first he wrote in English, 'I saw a couple of penguins at an empty nest today, in the midst of a group of occupied nests.'"; Ibid.

p. 205   *"I saw another astonishing sight of depravity today."*; Ibid.

p. 205   "As he put it so succinctly, 'There seems to be no crime too low for these penguins.'"; Ibid.

p. 207   *"Nothing I had experienced in the Ross Sea or in any other part of the world came up to the gales and blizzards of Commonwealth Bay"*; John King Davis, *High Latitude* (Melbourne: Melbourne University Press, 1962).

p. 207   "After a 'pleasant and uneventful trip,' according to Priestley"; Priestley, *Antarctic Adventure.*

p. 208   "'we would all have sworn that if there was one place along the coast which would be accessible in February, this would be the one.'"; Ibid.

p. 208   "causing the pack ice and bergs to 'bank up' on its southern side, and then to 'stream northwards' from its tip"; Ibid.

p. 209   *"during the past ages the Antarctic has possessed a climate much more genial than that of England at the present day"*; Ibid.

p. 210   "'This told us the whole story,' writes Scott that evening."; Scott, *Scott's Last Expedition.*

p. 210   *"Great God! this is an awful place and terrible enough for us to have laboured to it without the reward of priority"*; Ibid.

p. 210   *"I imagine it was intended to mark the exact spot of the Pole as near as the Norwegians could fix it."*; Ibid.

p. 211   *"Well, we have turned our back now on the goal of our ambition and must face our 800 miles of solid dragging"*; Ibid.

p. 211   "Bowers writes that they are 'thinning' and 'get hungrier daily'"; Cherry-Garrard, *The Worst Journey in the World.*

p. 211   *"God help us, with the tremendous summit journey and scant food."*; Scott, *Scott's Last Expedition.*

p. 214   *"The first of these is that the chick's downy coats become thick enough to protect them from cold"*; Levick, *Antarctic Penguins.*

p. 214   "The individual care of the chicks by their parents is abandoned, and in place of this, colonies start to 'pool' their offspring."; Ibid.

p. 216   *"The crimes which they commit are such as to find no place in this book"*; Ibid.

SEVENTEEN: WEATHER

p. 218   "Wilson, the doctor, records that Oates's 'big toe is turning blue-black.'"; Cherry-Garrard, *The Worst Journey in the World*.

p. 218   "'The weather is always uncomfortably cold and windy,' according to Wilson."; Ibid.

p. 218   "despite Scott noting that Evans 'is going steadily downhill,' the next day he allows Wilson to take rock samples"; Scott, *Scott's Last Expedition*.

p. 219   "'We cannot do distance without the ponies,' Scott tells his men."; Ibid., and Cherry-Garrard, *The Worst Journey in the World*.

p. 219   "with nothing but glaring white ahead, darkened by snow goggles, it is simply a form of mental starvation"; Levick, diaries, Scott Polar Research Institute.

p. 219   *"the way in which I as the chief character must avoid making such mistakes in the second."*; Ibid.

p. 220   "A considerable number of adults are still in full moult, and a few have finished moulting"; Ibid.

p. 220   *"driving them one by one into the water in response to the newly found instinct to catch their own food there."*; Ibid.

p. 220   "'A very terrible day,' as Scott observes."; Scott, *Scott's Last Expedition*.

p. 220   *"the remainder of us were forced to pull very hard, sweating heavily"*; Ibid.

p. 221   "He is 'on his knees with clothing disarranged, hands uncovered and frostbitten, and a wild look in his eyes.'"; Ibid.

p. 221   *"It is a terrible thing to lose a companion in this way"*; Ibid.

p. 221   "'Pray God,' he writes in his diary"; Ibid.

p. 221   *"February 17th: Still blowing hard with drift"*; Campbell, *The Wicked Mate*.

p. 222   "Levick's nose gets 'rather badly frostbitten.'"; Levick, diaries, Scott Polar Research Institute.

p. 222   *"We are now a little anxious about the ship, which was due on the 18th."*; Ibid.

p. 223   "Scott writes perceptively, 'It is a race between the season and hard conditions and our fitness and good food.'"; Scott, *Scott's Last Expedition*.

p. 223   *"There is little doubt we are in for a rotten critical time going home"*; Ibid.

p. 223   "They have scarcely enough, Scott realizes, even with the 'most rigid economy' to get them to their next depot"; Ibid.

p. 224   "Blizzards have kept them 'practically confined to our bags for 13 days—a record I believe for any antarctic party, and it has been absolutely miserable,'"; Levick, diaries, Scott Polar Research Institute.

p. 224   "for them, cruelly, to spy what they think is 'smoke on the horizon and under it a small black speck.'"; Campbell, *The Wicked Mate*.

p. 224   "Instead of their ship, it turns out to be 'only an iceberg with a cloud behind it.'"; Ibid.

p. 225   *"the road to hell might be paved with good intentions, but to us it seemed probable that hell itself would be paved something after the style of Inexpressible Island"*; Priestley, *Antarctic Adventure*.

p. 225   "Campbell's chances of relief are getting woefully small," writes Wilfred Bruce"; Hooper, *The Longest Winter*.

p. 226   "pack ice which they only manage to get out of 'with much difficulty.'"; Ibid.

p. 226   "'Our faces shone in rivalry with the sun,' Roald Amundsen, first man to the South Pole, will say"; Amundsen, *The South Pole*.

p. 227   *"I think of you and what you may wish, more than of him, and am in a strange mood"*; Huntford, *Scott and Amundsen*.

p. 227   "Nansen even goes on to say to her, "I wish that Scott had come first."; Ibid.

p. 229   "Scott notes fatalistically now, 'The dogs which would have been our salvation have evidently failed.'"; Scott, *Scott's Last Expedition*.

EIGHTEEN: DOGS

p. 231   *"I confess I had my misgivings. I had never driven one dog, let alone a team of them"*; Cherry-Garrard, *The Worst Journey in the World*.

p. 231   "asked that they increase the rations for the dogs because they were 'losing their coats.'"; Ibid.

p. 232   "'the chance of seeing another party at any distance was nil.'"; Ibid.

p. 232   "Ironically, they make '23 to 24 miles (statute) for the day.'"; Ibid.

p. 232   *"if he went under now, I doubt whether we could get through"*; Scott, *Scott's Last Expedition*.

p. 233   "Scott orders Wilson to 'hand over the means of ending our troubles.'"; Ibid.

p. 234   *"Abbott, Browning & I have killed & butchered 8 seals"*; Levick, diaries, Scott Polar Research Institute.

p. 234   "it is going to be a queer time for us through the dark months."; Ibid.

p. 234   *"when we ought to be getting on with the sealing and work on the cave, and we are losing the sun daily."*; Ibid.

p. 238   "According to Scott, 'We knew that poor Oates was walking to his death' and 'tried to dissuade him.'"; Scott, *Scott's Last Expedition*.

p. 238   *"I am just going outside and may be some time."*; Ibid.

p. 238   "remarkably, 'at Wilson's special request,' Scott consents to the remaining three of them continuing to pull the thirty-five pounds"; Ibid.

p. 241   "The wind increased to hurricane force, and suddenly one of the tent poles (on the lee side) broke with a snap"; Levick, diaries, Scott Polar Research Institute.

p. 241   "the pressing of the tent on them "produced a helpless suffocating sensation"; Ibid.

p. 241   "they are unable to find anywhere sheltered enough to give them 'the ghost of a chance' of erecting their spare tent"; Ibid.

p. 242   "'I shall always remember the appearance of Brownings (sic) face,' wrote Levick afterward"; Ibid.

p. 242   "to spend 'a most uncomfortable night' according to Campbell"; Campbell, *The Wicked Mate*.

p. 242   *"the prospect of the winter before us is enough to give anyone the hump I should think"*; Levick, diaries, Scott Polar Research Institute.

p. 242   "They 'have two days' food but barely a day's fuel'"; Scott, *Scott's Last Expedition*.

p. 242   "Scott's right foot is badly frostbitten: 'Amputation is the least I can hope for now.'"; Ibid.

p. 242   "the blizzard rages unabated and 'outside the door of the tent it remains a scene of whirling drift.'"; Ibid.

NINETEEN: WINTER

p. 244   "what Priestley describes as the 'visible darkness' afforded by the faint light"; Priestley, *Antarctic Adventure.*

p. 245   *"We are settling into our igloo now, and a dismal hole it is too."*; Levick, diaries, Scott Polar Research Institute.

p. 249   "'Campbell,' he replies."; Cherry-Garrard, *The Worst Journey in the World.*

p. 249   *"just then it seemed to me unthinkable that we should leave live men to search for those who were dead"*; Ibid.

p. 249   "Campbell's men 'might die for want of help.'"; Ibid.

p. 249   "'Were we to forsake men who might be alive to look for those whom we knew were dead?'"; Ibid.

p. 250   "Levick describes it as 'a great day of feasting.'"; Levick, diaries, Scott Polar Research Institute.

p. 250   *"I have not realised how hungry I have been during the last month or so"*; Ibid.

p. 250   "Priestley says, 'none of the famous wines of the world could possibly taste to us as did this,'"; Priestley, *Antarctic Adventure.*

p. 251   *"The hoosh flavoured with seal's brain and penguins's liver, was sublime"*; Ibid.

p. 252   "when they 'once more went back to a subnormal allowance' of food"; Ibid.

p. 252   "His fur mit (sic) was nearly full of blood which soon froze into a solid block."; Levick, diaries, Scott Polar Research Institute.

p. 252   "his hands are 'filthy & soaked with blubber from the stove,'"; Ibid.

p. 252   *"I shall feel rotten about it if his tendons are cut"*; Ibid.

p. 252   "He observes, 'The tendons of three fingers are cut I am sorry to say.'"; Ibid.

p. 252   "'We had another double hoosh,' Campbell notes"; Campbell, *The Wicked Mate.*

p. 254   "Levick, ever one to display his Victorian-bred roots, pronounces it 'a most boring production.'"; Levick, diaries, Scott Polar Research Institute.

p. 254   "This inevitably involves 'dining sumptuously at the various inns on the way,'"; Lambert, *'Hell with a Capital H.'*

p. 254   *"It is uncommonly cheering to think of the stretches of white dusty road at home at the present time"*; Levick, diaries, Scott Polar Research Institute.

p. 254   "down the Saskatchewan, and writing about it, with plenty of good photographs"; Ibid.

p. 254   *"Campbell & I spend hours over planning my trip down the Saskatchewan."*; Ibid.

TWENTY: RETURN JOURNEY

p. 255   *"Personally I am looking forward to the sledge journey before us with mixed feelings"*; Levick, diaries, Scott Polar Research Institute.

p. 256   *"Campbell says he means to start on or about the 22nd Sept."*; Ibid.

p. 256   *"The epidemic of diarrhoea continues in spite of precautions"*; Ibid.

p. 256   "Campbell 2. Levick 2. Priestley 4. Abbott 3. Browning 3. Dickason 2."; Ibid.

p. 256   "only to have Priestley come shuffling down the shaft 'in the last extremity.'"; Ibid.

p. 256   "Levick writes: '"Our small stack of literature is disappearing fast.'"; Ibid.

p. 257   "Midwinter's feast on June 22: 'One of the memorable days of our lives.'"; Ibid.

p. 258   "'I believe they thought we were ghosts,' wrote Levick afterward."; Ibid.

p. 259   "then walks up to Cherry-Garrard, saying simply, 'It is the tent.'"; Cherry-Garrard, *The Worst Journey in the World.*

p. 259   *"That scene can never leave my memory.";* Ibid.

p. 259   "There is a loud crack 'like a shot being fired'"; Sara Wheeler, *Cherry: A Life of Apsley Cherry-Garrard* (London: Jonathan Cape, 2001).

p. 260   *"It is the happiest day for nearly a year—almost the only happy one.";* Cherry-Garrard, *The Worst Journey in the World.*

p. 260   "'We were entirely free from fat, and, indeed, were so lean that our legs and arms were corrugated.'"; Lambert, *'Hell with a Capital H.'*

p. 261   *"There was no sound from behind except a faint, plaintive whine from one of the dogs";* Sir Douglas Mawson, *The Home of the Blizzard: Being the Story of the Australasian Antarctic Expedition, 1911–1914* (London: Ballantyne Press, 1915).

p. 261   "Mawson and Mertz rush back to discover 'a gaping hole in the surface about eleven feet wide.'"; Ibid.

p. 261   *"No sound came back but the moaning of a dog, caught on a shelf just visible one hundred and fifty feet below";* Ibid.

p. 261   "After some hours 'stricken dumb' and calling forlornly down the crevasse"; Ibid.

p. 266   "The Adélie penguin has a hard life: the Emperor penguin a horrible one."; Cherry-Garrard, *The Worst Journey in the World.*

p. 266   "With bright sunlight, a lop on the sea which splashed and gurgled under the ice-foot"; Ibid.

p. 267   "They make a 'thin soup' for themselves"; Mawson, *The Home of the Blizzard.*

p. 267   "is 'tough, stringy and without a vestige of fat'"; Ibid.

p. 267   *"On sledging journeys it is usual to apportion all food-stuffs in as nearly even halves as possible.";* Ibid.

p. 268   *"A long and wearisome night";* Ibid.

p. 268   "'this is terrible; I don't mind for myself but for others. I pray to God to help us.'"; Ibid.

p. 268   *"There appeared to be little hope of reaching the Hut."* Ibid.

p. 268   "He is shocked by what he finds: 'the thickened skin of the soles had separated in each case as a complete layer.'"; Ibid.

p. 270   "'Facing out over the Barrier, we gave three cheers and one more.'"; Cherry-Garrard, *The Worst Journey in the World.*

p. 272   *"Everything jet black & horribly greasy & smelling of blubber.";* Hooper, *The Longest Winter.*

PART FIVE: AFTER ANTARCTICA

TWENTY-ONE: THE DEPRAVITIES OF MEN

p. 278   *"The deity of success is a woman, and she insists on being won, not courted.";* Amundsen, *The South Pole.*

p. 281   *"We landed to find the Empire—almost the civilized world—in mourning.";* Cherry-Garrard, *The Worst Journey in the World.*

p. 281   *"Not here! The white South has thy bones";* *Daily Mail,* February 11, 1913, 4.

p. 282   "Amundsen, she wrote in her diary, 'looked unspeakably bored.'"; Kari Herbert, *Heart of the Hero: The Remarkable Women Who Inspired the Great Polar Explorers* (Glasgow: Saraband, 2013).

p. 282   *"Oh, well, never mind! I expected that.";* Lady Kathleen Kennet, *Self-Portrait of an Artist: From the Diaries and Memoirs of Lady Kennet, Kathleen, Lady Scott* (London: John Murray, 1949).

p. 282 "'Let me maintain a high, adoring exaltation, and not let the contamination of sorrow touch me.'"; Ibid.

p. 283 *"How awful if you don't."*; Herbert, *Heart of the Hero.*

p. 283 "addressed to her by Scott inasmuch as it says, 'To My Widow.' "; Ibid and partially in Scott, *Scott's Last Expedition.*

p. 284 *"I think the best chance has gone we have decided not to kill ourselves but to fight it to the last"*; Herbert, *Heart of the Hero.*

p. 289 "He found 'the ship packed above and below by a mass of unfortunate men, the majority severely wounded.'"; Murray Levick, letter to Mayson Beeton, The Keep, Sussex.

p. 289 *"Many of the wounds had not been dressed since they left the field and were crawling with maggots"*; Ibid.

p. 289 "he tells Beeton, 'refusing to return to my ship when they sent a boat for me.'"; Ibid.

p. 289 *"I was hoping that I would be court martialled, so that I could have an opportunity of making some sort of fuss"*; Ibid.

p. 289 "He notes that the wounded have been treated with 'most scandalous neglect'"; Ibid.

p. 289 "helping Levick 'by throwing amputated limbs over the side.'"; Henry R. Guly, "George Murray Levick (1876–1956), Antarctic explorer," *Journal of Medical Biography* 24, no. 1 (2014): 4–10.

p. 293 *"I know that during that long and racking march of thirty-six hours over the unnamed mountains and glaciers"*; Sir Ernest Shackleton, *South: The Story of Shackleton's 1914–1917 Expedition* (London: William Heinemann, 1919).

p. 294 "will always be able to enjoy it—certainly when you have forgotten all about it"; David Day, *Flaws in the Ice: In Search of Douglas Mawson* (Melbourne: Scribe, 2013).

p. 295 *"For scientific discovery, give me Scott; for speed and efficiency of travel, give me Amundsen"*; Raymond Priestley, 1956 address to the British Association for the Advancement of Science in which he paraphrased Cherry-Garrard introduction to *The Worst Journey in the World.*

TWENTY-TWO: AFTER THE WAR

p. 297 "He had retired from the navy the year after Gallipoli, furloughed on the 'grounds of unfitness.'"; Guly, "George Murray Levick (1876–1956), Antarctic explorer."

p. 299 "'How wrong that is,' she is reported to have said."; Michelle Merrilees, "Lady Shackleton: The Full Story," 2017, https://womenofeastbourne.co.uk/wp-content/uploads/2017/11/Lady-Shackleton-the-full-story.pdf.

p. 300 "By her own count, she will have 'three husbands and a hundred lovers.'"; Eric Utne, "Brenda, My Darling: The Love Letters of Fridtjof Nansen to Brenda Ueland," *Huffington Post*, March 10, 2012, https://www.huffpost.com/entry/great-love-letters_b_1192446.

p. 300 *"Here from my window in my tower, I see the maidenly birches in their bridal veils against the dark pine wood"*; Eric Utne, ed. *Brenda My Darling: The Love Letters of Fridtjof Nansen to Brenda Ueland* (Minneapolis: Utne Institute, 2011).

p. 303 *"Let me first fly to the Pole and back, and then we shall see."*; Tor Bomann-Larsen, *Roald Amundsen* (Stroud, UK: The History Press, 2011).

ENDNOTES

p. 305   *"We must acknowledge that in ascending the Barrier, Borchgrevink opened the way to the south"*; Amundsen, *The South Pole.*

p. 310   *"With very great regret & sympathy I have to inform you that your son was accidentally killed"*; Murray Levick, Expedition Notebooks, Archives, London: British Exploring Society.

p. 311   *"To Your Memory From Your Companions*: for fuck's sake Levick, is that the best you can do?"; Ibid.

p. 324   *"Color, which is the poet's wealth, is so expensive"*; Henry David Thoreau, Odell Shepard, ed., *The Heart of Thoreau's Journals* (New York: Dover Publications, Inc., 1961).

p. 324   *"Do I contradict myself?"*; Walt Whitman, *Leaves of Grass* (Philadelphia: David Mckay, 1892).

p. 325   *"And thus we sit together now"*; Robert Browning, *Dramatic Lyrics* (London: Browning, 1842).

p. 326   *"All the world loves a penguin: I think it is because in many respects they are like ourselves"*; Cherry-Garrard, *The Worst Journey in the World.*

INDEX